SpringerBriefs in Mathematics

SpringerBriefs in Mathematics showcases expositions in all areas of mathematics and applied mathematics. Manuscripts presenting new results or a single new result in a classical field, new field, or an emerging topic, applications, or bridges between new results and already published works, are encouraged. The series is intended for mathematicians and applied mathematicians.

More information about this series at http://www.springer.com/series/10030

Susana C. López • Francesc A. Muntaner-Batle

Graceful, Harmonious and Magic Type Labelings

Relations and Techniques

Springer

Susana C. López
Department of Mathematics
Universitat Politècnica de Catalunya
Castelldefels, Spain

Francesc A. Muntaner-Batle
School of Electrical Engineering
  and Computer Science
Faculty of Engineering and Built
  Environment
The University of Newcastle
Newcastle, NSW, Australia

ISSN 2191-8198   ISSN 2191-8201   (electronic)
SpringerBriefs in Mathematics
ISBN 978-3-319-52656-0   ISBN 978-3-319-52657-7   (eBook)
DOI 10.1007/978-3-319-52657-7

Library of Congress Control Number: 2017930429

Mathematics Subject Classification (2010): 05C78, 05C76, 05E99

Printed on acid-free paper

This Springer imprint is published by Springer Nature
The registered company is Springer International Publishing AG
The registered company address is: Gewerbestrasse 11, 6330 Cham, Switzerland

To our families.

Preface

The book that the reader has in her/his hands comes from the necessity of filling a gap in the graph labeling literature. Graph labeling is an exciting topic belonging to the area of graph theory that has gained a lot of popularity in the last years, and many new papers appear day after day. It is notorious that a subject with a dynamic survey being updated yearly, and over 2000 papers, counts only with four books,[1] one of them being a new edition of an existing one. With our book, we pretend modestly to contribute to the development of graph labeling, but at the same time being perfectly conscious that just a single book is not enough to cover all the material revealed by over 2000 papers. For sure, this area needs more books to be published in the future. Nevertheless, we feel that the existing material together with this work may be a starting point to increase the popularity of graph labeling among mathematicians.

This book covers different concepts related to some of the more popular labelings found in the literature, as well as the relations existing among them and some of the techniques used in the area. The main scope is to introduce the reader to the world of graph labeling from the point of view of the relations existing among labelings and the techniques used to understand them. Sometimes the relations established are techniques in their own right, and the same techniques apply to different labelings. These facts make it very difficult to divide the book into independent chapters and it is possible to find labelings of, let's say, type A in a chapter devoted to labelings of type B. But we believe that this is precisely one of the strongest points of this book, and it shows the strong connections among different labelings, which unfortunately have not been considered sufficiently by researchers in the past.

---

[1] – W.D. Wallis. Magic Graphs. Birkhäuser, Boston, 2001 (and a second ed. by A.M. Marr and W.D. Wallis, 2013);

– M. Bača and M. Miller. Super edge-antimagic graphs. BrownWalker Press, Boca Raton, 2008,
– M. Haviar, M. Ivaška. Vertex Labellings of Simple Graphs. Research and Exposition in Mathematics, Volume 34, Heldermann Verlag, 2015.

This book is addressed to a wide spectrum of readers that goes from anyone who wants to get introduced in the subject up to researchers in the field. However the reader is expected to have a šolid mathematical formation. Thus, we believe that the book is very good to be used in the last years of the mathematics major, for graduated students, and specially to serve as a guide for research seminar courses.

In Chap. 1, we develop some basic ideas of graph theory that will be specially useful to fix the notation. For readers not familiar with this subject we recommend to follow other classical books in graph theory, we provide some references there in. The second chapter is devoted to introduce the main labelings discussed through the coming chapters and many open problems are presented. It may be thought that is too soon to get into open problems, but we believe that the solution for many of these problems may come just from the insight of the problem solver, and probably not many techniques are needed. Therefore we feel that it is important for the reader to encounter these as soon as possible, and to get introduced to them with a fresh mind, to enforce the possibility to find new, inspiring and beautiful solutions for them. We conclude Chap. 2 with a section devoted to the origins of graph labelings, so that the reader can have a better idea of where this area is coming from. In Chap. 3 we established the first set of relations among labelings that are summarized in Fig. 3.2, and we provide an introduction to super edge-magic labelings since they are the main link for these relations. Chapters 4 and 5 are devoted to harmonious and graceful labelings respectively. Chapters 6 and 7 are fully devoted to study two different techniques that have been proven to be very useful when studying properties of labelings. The technique used in Chap. 6 is purely topological, while in Chap. 7, we concentrate in an algebraic method, originally introduced by David Hibert and then modified by Noga Alon. Both techniques are very powerful and many results have been obtained by using them.

Through the book, we have included a good deal of complete proofs, numerous exercises of different levels of difficulty and a good deal of open problems. We have combined classical results in the subject together with new techniques and results, that help the reader to obtain a solid base in graph labelings as well as to get an overview for some new research lines. The next scheme shows how the chapters are linked among them and provides alternative orders for reading the book.

$$\text{Chapter 1} \rightarrow \text{Chapter 2} \rightarrow \text{Chapter 3} \begin{array}{l} \nearrow \text{Chapter 4} \searrow \nearrow \text{Chapter 6} \\ \searrow \text{Chapter 5} \nearrow \searrow \text{Chapter 7} \end{array}$$

In order to conclude these lines we want to acknowledge the help and encouragement that we have obtained from our colleagues and friends Rikio Ichishima, Akito Oshima and Miquel Rius-Font, the very valuable and key information provided by Alex Rosa, as well as the comments made by the referees that have been very useful to increase the quality of the book. Many thanks to all of you.

Castelldefels, Spain   Susana C. López
Newcastle, NSW, Australia   Francesc A. Muntaner-Batle
September 2016

Acknowledgements

We are highly indebted with Dr. Anna Lladó for the excellent work performed by her in graph labeling and decomposition and in particular as the advisor of our thesis.

The second author wants to acknowledge Dr. Eduardo Canale-Bentancourt since the mathematical discussions maintened with him were crucial for the future development of the material found in Chap. 6.

The research conducted in this document by the first author has been supported by the Spanish Research Council under projects MTM2011-28800-C02-01, MTM2014-60127-P, by the Catalan Research Council under grant 2009SGR1387 and symbolically under grant 2014SGR1147.

Contents

Chapter 1
Notation and Terminology

1.1 Basic Terminology and Notation Concerning Graphs

This chapter contains notation and terminology used in the book. The remaining
concepts not found here and needed in the book will be defined in the corresponding
chapter.

The reader who is familiar with graphs can skip this chapter. For those who are
new to the field, we recommend other textbooks on graph theory, where the material
is presented with a lot of detail. For instance, we refer the reader to any of the
following sources [1–4, 6], which are excellent for anyone who wants to get any
contact with graph theory or those who want to get a deep knowledge of the subject.
A great number of references and possible topics of research can be found on their
pages, as well as a great deal of open problems.

1.1.1 Graphs, Subgraphs, and Isomorphism

Definition 1.1 A *simple graph* G is an ordered pair of sets $V(G)$ and $E(G)$ such
that the elements of $E(G)$ are unordered pairs of distinct elements of $V(G)$. The
elements of $V(G)$ are called the *vertices* and the elements of $E(G)$ the *edges*. We
write uv in order to denote the edge $\{u, v\}$. If $e = uv \in E(G)$, then u and v are
adjacent vertices, and we say that u and v are *incident* to e. We also say that u and v
are the *end-vertices* of e. Two edges are *incident* if they have a common end-vertex.
If $uv \notin E(G)$, then u and v are *independent*.

The *order* of G is $|V(G)|$ and the *size* is $|E(G)|$, where $|\cdot|$ denotes the cardinality
of the set. In the book, we only consider graphs with finite order and size. We say
that G is a simple (p, q)-*graph* if the order of G is p and its size is q.

© The Author(s) 2017   1
S.C. López, F.A. Muntaner-Batle, *Graceful, Harmonious and Magic Type Labelings*,
SpringerBriefs in Mathematics, DOI 10.1007/978-3-319-52657-7_1

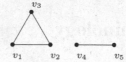

Fig. 1.1 A drawing of the graph G

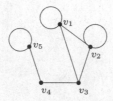

Fig. 1.2 A $(5, 8)$-graph with at most one loop attached at each vertex G

It is common to represent graphs as drawings on the plane in which the vertices are represented by dots and the edges are represented by curves joining the vertices that represent each edge.

Example 1.1 Figure 1.1 shows a possible drawing for the simple $(5, 4)$-graph G defined by $V(G) = \{v_1, v_2, \ldots, v_5\}$ and $E(G) = \{v_1 v_2, v_1 v_3, v_2 v_3, v_4 v_5\}$.

Occasionally, we admit edges with identical end-vertices, that are called *loops*, and repeated edges, that are called *multiple edges*. The term simple graph explicitly forbids multiple edges and loops, while the term *multigraph* explicitly allows them. We use the term *graph* to denote a simple graph and, in Chap. 6, simple graphs with at most one loop attached at each vertex.

Example 1.2 Figure 1.2 shows a possible drawing for the $(5, 8)$-graph G defined by $V(G) = \{v_1, v_2, \ldots, v_5\}$ and $E(G) = \{v_1 v_1, v_1 v_2, v_1 v_3, v_2 v_2, v_2 v_3, v_3 v_4, v_4 v_5, v_5 v_5\}$.

Definition 1.2 We say that H is a *subgraph* of G, and we denote it by $H \subseteq G$, when $V(H) \subseteq V(G)$ and $E(H) \subseteq E(G)$. If H contains all edges of G with end-vertices in $V(H)$, we say that H is an *induced subgraph*. If H is an induced subgraph of G with set of vertices S, we denote it by $H = G[S]$ and we say that H is the subgraph induced by S.

Some particular types of subgraphs are the following ones. A *spanning subgraph* of G is a subgraph H with $V(H) = V(G)$. Suppose that $S \subseteq V(G)$. Then $G - S$ denotes the subgraph $G[V(G) \setminus S]$. Similarly, if $T \subseteq E(G)$, then $G - T$ denotes the subgraph with vertex set $V(G)$ and edge set $E(G) \setminus T$. When $S = \{u\}$ and $T = \{uv\}$ we simply write $G - u$ and $G - uv$.

Let $u, v \in V(G)$. If $uv \notin E(G)$, then $G + uv$ denotes the graph obtained from G by adding the edge uv.

Example 1.3 Let G be the graph in Fig. 1.3, and let $S = \{v_5, v_6, v_7\}$ and $T = \{v_1 v_2, v_1 v_8, v_5 v_8\}$. The subgraphs $G - S$ and $G - T$ are shown in Fig. 1.3.

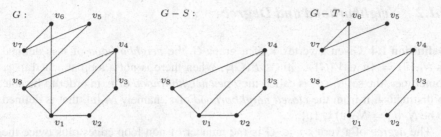

Fig. 1.3 A graph G and the subgraphs $G - S$ and $G - T$

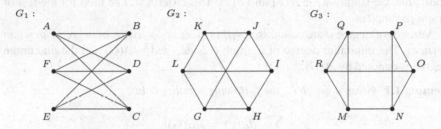

Fig. 1.4 The graphs G_1, G_2 and G_3

Then $H \subseteq G$ defined by $V(H) = \{v_1, v_2, v_3, v_4\}$ and $E(H) = \{v_1v_2, v_2v_3, v_3v_4\}$ is a subgraph of G which is not induced by any subset of vertices. On the other hand, the subgraph $H' = H + v_1v_4$ is the subgraph induced by $\{v_1, v_2, v_3, v_4\}$. Thus, $H' = G[\{v_1, v_2, v_3, v_4\}]$. The subgraphs H and H' of G are not spanning. However, $G - T$ is a spanning subgraph of G.

One of the questions that raises up when studying graph theory is when we can consider two graphs to be equal. In fact, there are several degrees of equality between graphs. This is explained using the concepts of isomorphism and automorphism between graphs. However, to distinguish between isomorphic and non isomorphic graphs is not always an easy task.

Definition 1.3 Given two graphs G and H we say that G and H are *isomorphic*, written $G \cong H$, if there is a bijective function $\phi : V(G) \to V(H)$ that preserves the adjacency relation. That is, there is an edge $uv \in E(G)$ if and only if, $\phi(u)\phi(v)$ is an edge in H. Such a function ϕ is called an *isomorphism* from G to H.

An isomorphism from a graph to itself is called an *automorphism*.

Exercise 1.1 Show that $G_1 \cong G_2$ and that $G_2 \not\cong G_3$, where G_1, G_2, and G_3 are the graphs shown in Fig. 1.4.

Exercise 1.2 Show that the relation graph G is related to graph H if and only if $G \cong H$ is an equivalence relation on the set of graphs.

1.1.2 Neighborhood and Degree

Definition 1.4 Given a vertex v of a graph G, the *neighborhood* of v is defined as $N_G(v) = \{u \in V(G) : uv \in E(G)\}$. When there is not a loop attached to v, sometimes, the set $N_G(v)$ is called the *open neighborhood* of v, in order to be able to distinguish it from the *closed neighborhood* of v, namely $N_G[v]$, that is defined to be $N_G[v] = N_G(v) \cup \{v\}$.

The *degree* of a vertex $v \in G$ is the number of non-loop edges plus twice the number of loops incident with v, and it is denoted by $d_G(v)$. If there is no possible confusion, we simply write $N(v)$ and $d(v)$. This criteria will be used for the rest of graph parameters.

Vertices of degree 0 are called *isolated vertices* and vertices of degree 1 *pendant vertices*. The minimum degree of a graph G is denoted by $\delta(G)$ and the maximum degree is denoted by $\Delta(G)$.

Lemma 1.1 *Given a graph G, the following equality holds:*

$$\sum_{v \in V(G)} d_G(v) = 2|E(G)|.$$

Exercise 1.3 Prove Lemma 1.1.

Odd vertices refer to vertices with odd degree, and *even vertices* to vertices with even degree. Using Lemma 1.1, the following exercise can be easily solved.

Exercise 1.4 Prove that every graph has an even number of odd vertices.

1.1.3 Path, Components, Distance, and Diameter

Definition 1.5 A *path P* of length n in a graph G is a sequence of distinct vertices $v_0, v_1, \ldots, v_n$ in G where $v_i v_{i+1}$ is an edge for every $i = 0, 1, \ldots, n - 1$. We also say that P is a $v_0 v_n$-path.

Definition 1.6 A graph G is *connected* if it has a uv-path for each pair of vertices u and v; otherwise it is *disconnected*. A set of vertices or edges S is a *separating set* of a connected graph G if $G - S$ is disconnected. If G is a disconnected graph, then the connected subgraphs that are maximal with respect the inclusion relation are called the *connected components* or simply *components*. A vertex is a *cut-vertex* or an edge is a *cut-edge* (or a *bridge*) if its deletion increases the number of components of G. If every component of a disconnected graph consists of a single vertex, the graph is said to be *totally disconnected*. The totally disconnected graph on n vertices is denoted by N_n, and it is also known as the *null graph* of order n.

Example 1.4 Figure 1.1 shows a disconnected graph G with two connected components, namely $G[\{v_1, v_2, v_3\}]$ and $G[\{v_4, v_5\}]$. All graphs that appear in Fig. 1.3 are connected. Consider the graph $G - T$ in Fig. 1.3, we see that v_6 is a cut-vertex, $v_6 v_7$ is a cut-edge, and $\{v_1, v_8\}$ is a separating set of vertices of $G - T$.

Exercise 1.5 Show that $\mathscr{R}$, the relation on $V(G)$, defined by $u\mathscr{R}v \Leftrightarrow u$ and v belong to the same component is an equivalence relation.

Definition 1.7 Given two vertices u and v of G, the *distance* between vertices u and v, denoted by $d_G(u, v)$, is the minimum length of all uv-paths. If G does not contain a uv-path, we say $d_G(u, v) = +\infty$. Such a distance defines a metric in the set $V(G)$, in the proper sense, that is, for every $u, v, w \in V(G)$

- $d_G(u, v) \geq 0$, where equality holds if and only if $u = v$,
- $d_G(u, v) = d_G(v, u)$, and
- $d_G(u, v) \leq d_G(u, w) + d_G(w, v)$.

Example 1.5 Let G be the graph that appears in Fig. 1.3. The distances involving v_1 and v_3 are the following: $d(v_1, v_1) = d(v_3, v_3) = 0$, $d(v_1, v_2) = d(v_1, v_4) = d(v_1, v_6) = d(v_1, v_8) = 1$, $d(v_1, v_3) = d(v_1, v_5) = d(v_1, v_7) = 2$, $d(v_2, v_3) = d(v_3, v_4) = 1$, $d(v_3, v_8) = 2$, $d(v_3, v_5) = d(v_3, v_6) = 3$, and $d(v_3, v_7) = 4$.

Definition 1.8 The *diameter* of a connected graph G, denoted by $D(G)$, is the maximum distance between any pair of vertices of G. The *eccentricity* of a vertex $v \in V(G)$, denoted by $e(v)$, is the maximum distance between v and any other vertex. Thus, the diameter can be defined as $D(G) = \max_{v \in V(G)}\{e(v)\}$. The *radius* of G, denoted by $r(G)$, is $r(G) = \min_{v \in V(G)}\{e(v)\}$. The *center* of G is the set of vertices with eccentricity equal to the radius of the graph. Every element in this set is called a *central* vertex.

Example 1.6 According to Example 1.5, the graph G in Fig. 1.3 has radius $r(G) = 2$ and diameter $D(G) = 4$. The vertex v_1 is a central vertex.

1.1.4 Walk, Trail, Circuit and Cycle

A path in a graph is a particular case of a walk, that we introduce next.

Definition 1.9 A *walk* W in a graph G of length n is an alternating sequence of vertices and edges, namely, $v_0, e_1, v_1, e_2, \ldots, e_n, v_n$, where $e_i = v_{i-1}v_i$, for each $i \in \{1, 2, \ldots, n\}$. The vertices v_0 and v_n are the *end-vertices* of W. We also say that W is a $v_0 v_n$-walk. Thus, a path is a walk with distinct vertices. If all edges of a walk are distinct, then the walk is said to be a *trail* . A trail in which the end-vertices coincide is called a *circuit*. A *Eulerian (trail) circuit* is a (trail) circuit that contains all edges of the graph.

Definition 1.10 A graph is *Eulerian* if it has a Eulerian circuit.

Exercise 1.6 Prove that a non-trivial connected graph is Eulerian if and only if each vertex has even degree.

Definition 1.11 A *cycle* of length n in a graph G is a path $v_0, v_1, \ldots, v_n$ in G, with the additional property that $v_0 = v_n$. An *odd cycle* refers to a cycle of odd length and an *even cycle*, to a cycle of even length. The *girth* of a graph G is the length of the shortest cycle in G and the *circumference* is the length of the longest one.

Exercise 1.7 Prove that a non-trivial graph is Eulerian if and only there exists a partition $E(G) = \cup_{i \in J} T_i$ such that $G[T_i]$ is a cycle, for every $i \in J$.

1.1.5 Adjacency Matrix and Kronecker Product of Matrices

The representation of graphs using drawings is very common and useful, but it is not the only one. Another useful representation of a graph is by means of the adjacency matrix.

Definition 1.12 The *adjacency matrix* of a (multi)graph G, with vertices indexed as $V(G) = \{v_1, v_2, \ldots, v_n\}$, denoted by $A(G)$, is the integer matrix in which a_{ij} is the number of copies of the edge $v_i v_j$.

In a simple graph, or in a graph in which at most one loop is allowed per vertex, every entry in the adjacency matrix is either 0 or 1.

There exist several ways to combine matrices. One of such ways is by using the Kronecker product of matrices.

Definition 1.13 If A is an $m \times n$ matrix and B is a $p \times q$ matrix, then the *Kronecker product* $A \otimes B$ is the $mp \times nq$ block matrix, obtaining by multiplying each element a_{ij} of A times B:

$$A \otimes B = \begin{pmatrix} a_{11}B & \cdots & a_{1n}B \\ \vdots & \ddots & \vdots \\ a_{m1}B & \cdots & a_{mn}B \end{pmatrix}.$$

1.1.6 Some Well Known Graphs

There are several families of graphs that have received special attention in the area of graph labelings, since as we will see further in the book, many authors have concentrated their efforts to study whether a given family of graphs admits one labeling or another, even defining new families of graphs for this purpose. This is the reason why we devote this section to introduce some of such families.

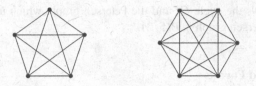

Fig. 1.5 The graphs K_5 and K_6

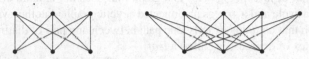

Fig. 1.6 The graphs $K_{3,3}$ and $K_{3,5}$

1.1.6.1   Complete Graphs

A simple graph G is the *complete graph* on p vertices, and it is denoted by K_p, if its order is p and its size is $q = \binom{p}{2}$. That is to say, all possible edges appear in $E(K_p)$. Figure 1.5 shows K_5 and K_6.

1.1.6.2   Bipartite Graphs

A graph G is said to be *bipartite* if it is possible to partition the set $V(G)$ into two sets V_1 and V_2, that we will call *stable sets*, such that every edge is incident to a vertex of V_1 and to a vertex of V_2. A bipartite simple graph, where $|V_1| = m$ and $|V_2| = n$, is called a *complete bipartite graph* and it is denoted by $K_{m,n}$, if its size is mn. That is, if each vertex in V_1 is adjacent to all vertices in V_2 and each vertex in V_2 is adjacent to all vertices in V_1. In Fig. 1.6, we show the graphs $K_{3,3}$ and $K_{3,5}$.

The following result is very useful when dealing with bipartite graphs.

Theorem 1.1 *A graph is bipartite if and only if it does not contain any odd cycle.*

Exercise 1.8 Prove Theorem 1.1.

1.1.6.3   Regular Graphs and Cycles

A graph is *r-regular* if $\delta(G) = \Delta(G) = r$. A 1-regular graph is called a *matching*, and if a graph G contains a *spanning matching* M, then M is called a *perfect matching*. A *cycle* is a connected 2-regular simple graph. Usually the cycle on n vertices, $n > 2$, is denoted by C_n. A 3-regular graph is usually called a *cubic graph* and one of the most important cubic graphs is the *Petersen graph*. In fact, a family of graphs that has been considered a lot in the world of graph labelings is the family of generalized Petersen graphs denoted by $P(n, k)$. The *generalized Petersen graph* [5] $P(n, k)$ has vertex set $V(P(n, k)) = \{a_i, b_i\}_{i=0}^{n-1}$ and edge set $E(P(n, k)) = \{a_i a_{i+1} : i \in \mathbb{Z}_n\} \cup \{a_i b_i : i \in \mathbb{Z}_n\} \cup \{b_i b_{i+k} : i \in \mathbb{Z}_n$ and the sum is taken in $\mathbb{Z}_n\}$.

Figure 1.7 shows the cycle C_5^- and the Petersen graph, which using the notation of generalized Petersen graph is $P(5, 2)$.

1.1.6.4   Trees and Forests

An *acyclic graph* or *forest* (Fig. 1.8) is a graph that does not contain any subgraph isomorphic to a cycle and a *tree* is a connected acyclic graph. Another way to define a tree is a graph that contains exactly one path between any pair of distinct vertices. In a tree, a vertex of degree 1 is called a *leaf*.

Every tree can be represented as a *rooted tree* T with a distinguished vertex r, called the *root*. Let $P(v)$ be the unique rv-path. Then, the only vertex in $N_{P(v)}(v)$ is called the *parent* of v and the *children* of v are its other neighbors in T.

Exercise 1.9   Show that a graph is a tree if and only if there is exactly a unique path between any pair of distinct vertices of the tree.

Exercise 1.10   Show that the center of a tree consists of either one vertex or two adjacent vertices (Fig. 1.9).

Exercise 1.11   Show that any tree with at least two vertices contains at least two leaves.

There are several classes of acyclic graphs that frequently appear in graph theory, and more particularly related to the graph labeling literature.

A *path* is a graph obtained from the deletion of any edge of a cycle and a *linear forest* is a forest for which each component is a path. Usually, a path with n vertices is denoted by P_n. Notice that, for paths and cycles, we have three possible meanings: as a graph or subgraph, as a special configuration of vertices an edges, and as a set of edges.

Fig. 1.7   The cycle C_5 and the Petersen graph $P(5, 2)$

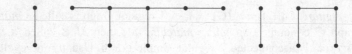

Fig. 1.8   A forest with four components

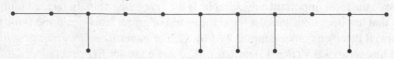

Fig. 1.9   A tree with exactly one vertex in its center

A *caterpillar* is a tree for which the deletion of its leaves results in a path, which is called the *spine* of the caterpillar. A *lobster* is any tree for which the deletion of its leaves produces a caterpillar. A *star* is any graph isomorphic to $K_{1,n}$, $n > 0$ (Fig. 1.10).

1.1.6.5 Unyciclic Graphs

A *unicyclic* graph is a graph that is not acyclic and there is at least one edge of the graph such that the deletion of this edge results in an acyclic graph (Fig. 1.11).

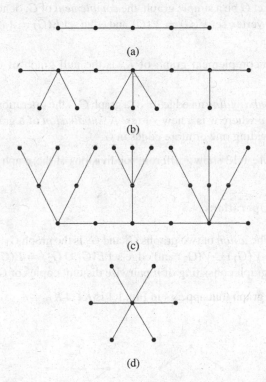

Fig. 1.10 (**a**) The path P_5, (**b**) a caterpillar, (**c**) a lobster, and (**d**) the star $K_{1,6}$

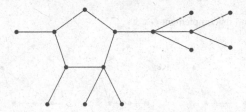

Fig. 1.11 A unicyclic graph

1.1.7   Graph Operations

There are different ways to obtain new graphs from old ones. Some of them are the result of different types of graph operations. In fact, there exist families of graphs in which the most convenient way to describe them is using these operations. Here we describe the most common operations on graphs.

1.1.7.1   Unary Operations

Definition 1.14 Let G be a simple graph, the *complement* of G, denoted by $\overline{G}$, is the simple graph with vertex set $V(\overline{G}) = V(G)$ and edge set $E(\overline{G}) = \{uv : uv \notin E(G)\}$ (Fig. 1.12).

Notice that the complement graph of K_p is the null graph of order p, that is, $\overline{K_p} = N_p$.

Definition 1.15 *Subdividing* an edge uv of a graph G is the operation of substituting uv by a path u, w, v where w is a new vertex. A *subdivision* of a graph G is a graph obtained by subdividing one or more edges in G.

Example 1.7 Figure 1.13 shows different subdivisions of the graph G in Fig. 1.12.

1.1.7.2   Binary Operations

Definition 1.16 The *union* of two graphs G_1 and G_2 is the graph $G_1 \cup G_2$ with vertex set $V(G_1 \cup G_2) = V(G_1) \cup V(G_2)$ and edge set $E(G_1 \cup G_2) = E(G_1) \cup E(G_2)$. We denote by nG the graph consisting of n pairwise disjoint copies of G.

Example 1.8 The graph that appears in Fig. 1.1 is $K_3 \cup K_2$.

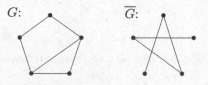

Fig. 1.12 A graph G and its complement $\overline{G}$

Fig. 1.13 Different subdivisions of a graph

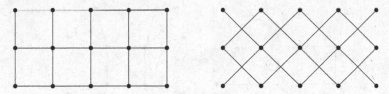

Fig. 1.14 The graph $P_5 \square P_3$, on the *right*, and the graph $P_5 \otimes P_3$, on the *left*

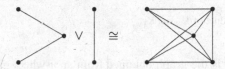

Fig. 1.15 The graph $P_3 \vee K_2$

Definition 1.17 The *Cartesian product* of two graphs G_1 and G_2 is the graph $G_1 \square G_2$ with vertex set $V(G_1) \times V(G_2) = \{(x_1, x_2) : x_1 \in V(G_1), x_2 \in V(G_2)\}$ and for which two vertices (x_1, x_2) and (y_1, y_2) are adjacent whenever $x_i y_i \in E(G_i)$ and $x_j = y_j$, for $j \neq i$.

Definition 1.18 The *direct product* of two graphs G_1 and G_2 is the graph $G_1 \otimes G_2$[1] with vertex set $V(G_1) \times V(G_2) = \{(x_1, x_2) : x_1 \in V(G_1), x_2 \in V(G_2)\}$ and for which two vertices (x_1, x_2) and (y_1, y_2) are adjacent whenever $x_i y_i \in E(G_i)$, for every $i = 1, 2$.

Example 1.9 The graphs $P_5 \square P_3$ and $P_5 \otimes P_3$ are drawn in Fig. 1.14.

Some graphs that are obtained using the Cartesian product are the following: the *book* B_n is obtained by the Cartesian product of two stars: $B_n = K_{1,n} \square K_2$; the *grid* $G_{m \times n}$ is the graph obtained by the Cartesian product $P_m \square P_n$; the *n-ladder* L_n is the graph obtained by the Cartesian product $K_2 \square P_n$; the *prism* Y_n is the Cartesian product $Y_n = K_2 \square C_n$; a *generalized prism* $Y_{m,n}$ is the graph obtained by the Cartesian product $Y_{m,n} = C_m \square P_n$.

Definition 1.19 The *join product* of two graphs G_1 and G_2 is the graph $G = G_1 \vee G_2$ with vertex set $V(G) = V(G_1) \cup V(G_2)$ and edge set $E(G) = E(G_1) \cup E(G_2) \cup \{xy : x \in V(G_1), y \in V(G_2)\}$.

Figure 1.15 shows the graph $P_3 \vee K_2$.

Some graphs that are obtained using the join product are the following ones: a *fan* F_n is the graph obtained by the join product of a path of length $n - 1$ with a single vertex, that is $F_n = P_n \vee K_1$; a *n-wheel* W_n is the graph obtained by the join product of a cycle of length n with a single vertex. Some authors use W_n to refer to a wheel graph of order n.

---

[1]Another standard notation for the direct product is $G_1 \times G_2$.

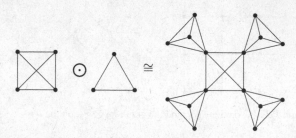

Fig. 1.16 The corona product of K_4 and K_3

The *helm* graph H_n is the graph obtained from an n-wheel graph by adjoining a pendant edge at each vertex of the cycle.

Definition 1.20 The *corona product* of two graphs G_1 and G_2 is the graph $G_1 \odot G_2$ obtained by placing a copy of G_1 and $|V(G_1)|$ copies of G_2 and then joining each vertex of G_1 with all vertices in one copy of G_2 in such a way that all vertices in the same copy of G_2 are joined with exactly one vertex of G_1.

Figure 1.16 shows the graph $K_4 \odot K_3$.

In order to conclude this section, we introduce the important concept of graph decomposition. As we will see, there is a strong connection between graph labelings and graph decompositions. In fact, graph decompositions have motivated what have been probably the most important and most studied labelings of all, namely graceful labelings. They were the first labelings to appear and we devote Chap. 5 to their discussion.

Definition 1.21 A *decomposition* of a graph G is a collection $\{H_i\}_{i=1}^r$ of subgraphs of G such that $\{E(H_i)\}_{i=1}^r$ is a partition of $E(G)$. If $\{H_i\}_{i=1}^r$ is a decomposition of G, we write:

$$G \cong H_1 \oplus H_2 \oplus \cdots \oplus H_r = \oplus_{i=1}^r H_i.$$

1.2   Basic Terminology and Notation Concerning Digraphs

A variation of the concept of a graph that will be of great use for us is the concept of directed graph or digraph.

Definition 1.22 A *(simple) directed graph* or *(simple) digraph* D is an ordered pair of sets $V(D)$ and $E(D)$, containing *vertices* and *arcs*, respectively, in which every arc is an ordered pair of (distinct) vertices, namely $a = (u, v)$. We say that u is the *tail* of a and v the *head* of a. We also say that u is *adjacent to* v or that v is *adjacent from* u. The *order* and *size* is defined similarly, as in the case of simple graphs. The notion of *subdigraph* extends easily by replacing edges by arcs.

An *oriented graph* is a digraph obtained from a graph G by replacing each edge uv by exactly one of the corresponding arcs, either (u, v) or (v, u). Conversely, the *underlying graph* obtained from a digraph D, denoted by $und(D)$, is the graph obtained from D by replacing each arc (u, v) with the corresponding edge uv.

In case of digraphs, we define the sets $N^+(v) = \{u \in V(D) : (u, v) \in E(D)\}$ and $N^-(v) = \{u \in V(D) : (v, u) \in E(D)\}$. Similarly, we define the *out-degree* $d^+(v)$ and the *in-degree* of a vertex $d^-(v)$ as the number of arcs that contain v as a tail and as a head, respectively, and we use the symbols $\delta^+(G)$, $\Delta^+(G)$, $\delta^-(G)$, and $\Delta^+(G)$ in order to denote the minimum and maximum out-degrees and in-degrees, respectively.

Exercise 1.12 Obtain the directed version of Lemma 1.1 for simple digraphs.

The concept of adjacency matrix of a graph extends easily for digraphs as follows.

Definition 1.23 The *adjacency matrix* of a digraph G, with vertices indexed as $V(G) = \{v_1, v_2, \ldots, v_n\}$, denoted by $A(G)$, is the integer matrix in which a_{ij} is the number of copies of the arc (v_i, v_j).

In a simple digraph, every entry in the adjacency matrix is either 0 or 1. Even when the underlying graph contains at most one loop attached at each edge, every entry of the adjacency matrix of the digraph is either 0 or 1.

Definition 1.24 A *walk* in a digraph is a sequence of vertices and arcs $v_0, e_1, v_1, e_2, \ldots, e_n, v_n$ such that either $e_i = (v_{i-1}, v_i)$ or $e_i = (v_i, v_{i-1})$ for each $i \in \{1, 2, \ldots, n\}$. The walk is *directed* if $e_i = (v_{i-1}, v_i)$, for each $i \in \{1, 2, \ldots, n\}$. Thus, the notions of *(directed) paths* and *cycles* extend easily.

A digraph is said to be *strongly connected* if for for each pair of vertices, u and v, there is a directed walk from u to v and a directed walk from v to u. The maximal strongly connected subdigraphs are called the *strong components*.

A digraph is *connected* if for any two vertices u and v there is a walk from u to v. That is, if its underlying graph is connected. A maximal connected subdigraph is called a *component*.

References

1. Bondy, J.A., Murty, U.S.R.: Graph Theory with Applications. American Elsevier Publishing Co., New York (1976)
2. Bondy, J.A., Murty, U.S.R.: Graph Theory, Graduate Texts in Mathematics, vol. 244. Springer, New York (2008)
3. Chartrand, G., Lesniak, L.: Graphs and Digraphs, 2nd edn. Wadsworth & Brooks/Cole Advanced Books and Software, Monterey (1986)
4. Hammarck, R., Imrich, W., Klavžar, S.: Handbook of Product Graphs, 2nd edn. CRC Press, Boca Raton, FL (2011)
5. Watkins, M.E.: A theorem on Tait colorings with an application to generalized Petersen graphs. J. Comb. Theory **6**, 152–164 (1969)
6. West, D.B.: Introduction to Graph Theory. Prentice Hall, Upper Saddle River, NJ (1996)

Chapter 2
Graphs Labelings

2.1 Basic Definitions

By a *labeling* of a graph, also known as a valuation of a graph, we mean a map that carries graph elements onto numbers (usually the positive or nonnegative integers) called *labels* that meet some properties depending on the type of labeling that we are considering. The most common choices for the domain are the set of vertices alone (*vertex labelings*), or edges alone (*edge labelings*), or the set of edges and vertices together (*total labelings*) (see [12]). Other domains are also possible, but they will not be discussed in this book.

In what follows we include the definitions of the more popular labelings together with examples and open problems. However, this book does not pretend to be a survey on graph labelings. For the reader interested in a very complete guide on graph labeling problems and results, we suggest [12]. The definitions needed in this book that are not introduced in this chapter will be presented as needed.

For integers $m \leq n$, we denote the set $\{m, m + 1, \ldots, n\}$ by $[m, n]$.

2.1.1 α and β Valuations

Rosa [31] introduced in 1967 the very first types of graph labelings or valuations (α, β, γ, and ρ valuations) as tools to solve a very famous conjecture on graph theory due to Ringel [28]. The conjecture states that every tree T of size n decomposes the complete graph K_{2n+1} into isomorphic copies.

We give the definitions of α and β valuations next.

Definition 2.1 ([31]) Let G be a (p, q)-graph and let $f : V(G) \to [0, q]$ be an injective function such that when each edge uv is assigned the value $g(uv) = |f(u) - f(v)|$, the resulting edge labels are all distinct. Then f is a β-*valuation* of G.

© The Author(s) 2017
S.C. López, F.A. Muntaner-Batle, *Graceful, Harmonious and Magic Type Labelings*,
SpringerBriefs in Mathematics, DOI 10.1007/978-3-319-52657-7_2

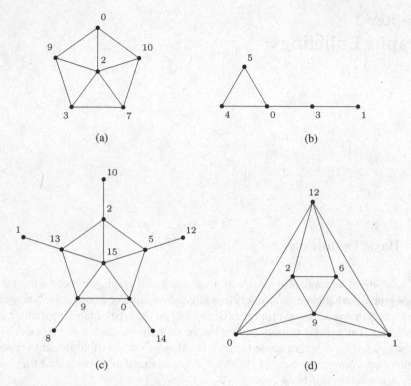

Fig. 2.1 Some examples of graceful labeled graphs

The β-valuation was renamed *graceful* by Golomb [13] and a graph that admits a graceful labeling is called a graceful graph. We mention that the term graceful labeling has become more popular than the original one. Although an entire chapter will be devoted to graceful labelings, we introduce in Fig. 2.1 some examples of graceful graphs with a corresponding graceful labeling. We also propose some open problems in the subject, although we leave the most famous conjecture on graceful labelings (and probably in the world of graph labelings) for Chap. 5.

Before getting started with the open problems, we introduce the following special case of graceful labelings called α-labelings.

Definition 2.2 ([31]) A graceful labeling f of a (p, q)-graph G is said to be an α-*valuation* of G if there exists an integer k with $0 \le k < q$, called the *characteristic* of f, such that $\min\{f(u), f(v)\} \le k < \max\{f(u), f(v)\}$, for every edge uv of G.

α-valuations are also named α-*labelings* .

An interesting question that remains open about graceful labelings is the following one.

Problem 2.1 ([27]) Characterize the set of graceful generalized Petersen graphs.

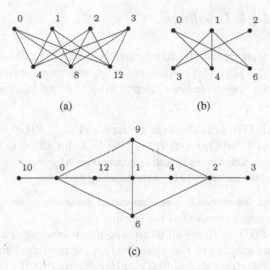

Fig. 2.2 Some examples of α-labeled graphs. [(c) Reprinted from S. C. López, F. A. Muntaner-Batle, A new application of the $\otimes_h$-product to α-labelings, Discrete Math. 338: 839–843, ©2015, with permission from Elsevier]

As we will see further in the book, a long-standing conjecture due to Truszczyński [36] states that all unicyclic graphs except for C_n, where $n \equiv 1$ or 2 (mod 4) are graceful. However this conjecture leads to the following open problem that seems to be extremely difficult.

Problem 2.2 Characterize the set of graceful graphs of order equal to size.

Other particular cases of the previous open problem can be stated as follows.

Problem 2.3 Although some families of 2-regular graphs are known either to admit or not to admit graceful labelings, for instance cycles (see Chap. 5), there are many others that are not known. To study the graceful properties of such families would also be of interest.

Some examples of α-labeled graphs are shown in Fig. 2.2. For many types of labelings it is common to find conjectures that state that all trees admit such a labeling. However, α-labelings are an exception to this rule, since it is not hard to find examples of trees that do not admit α-labelings. In spite of this, it is possible to find conjectures involving α-labelings and trees. Among them we want to point out the following one introduced in [5].

Conjecture 2.1 ([5]) All trees with maximum degree 3 and a perfect matching have an α-labeling.

As in the case of graceful labelings, it is unknown which 2-regular graphs admit α-labelings. Hence, we have the following open question.

Problem 2.4 ([19, 20, 31]) Characterize the 2-regular graphs that admit α-labelings.

2.1.2 Edge-Magic Labelings

Magic valuations were defined by Kotzig and Rosa in [21] and later rediscovered by Ringel and Lladó [29] who coined one of the now popular terms, edge-magic (EM). More recently, they have been referred to as EM total labelings by Marr and Wallis [24].

Definition 2.3 ([21, 29]) Let G be a (p, q)-graph and let $f : V(G) \cup E(G) \to [1, p + q]$ be a bijection such that $f(u) + f(uv) + f(v) = k$, for all $uv \in E(G)$. Then f is called an *edge-magic labeling* of G and G is called an *edge-magic graph* (EM). The constant k is called the *valence* [21] of the labeling f.

In the literature, the constant k is also called *the magic sum* [24], the *magic weight* [2], or *the magic constant* of the labeling f.

Let $f : V(G) \cup E(G) \to [1, p + q]$ be an edge-magic labeling of a (p, q)-graph G. The *complementary labeling* of f, denoted by $\bar{f}$, is the labeling defined by the rule: $\bar{f}(x) = p + q + 1 - f(x)$, for all $x \in V(G) \cup E(G)$. Notice that, if f is an edge-magic labeling of G with magic sum k, we have that $\bar{f}$ is also an edge-magic labeling of G with magic sum $\bar{k} = 3(p + q + 1) - k$.

Figure 2.3 shows two examples of edge-magic labelings of K_5 and C_4. Observe that the two labelings of K_5 are complementary of each other, as well as the labelings of C_4.

A long-standing conjecture on edge-magic graphs, and probably one of the hardest ones in the world of graph labelings is the edge-magic tree conjecture which appeared as an open problem in [21].

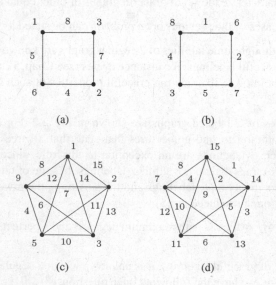

(a)   (b)

(c)   (d)

Fig. 2.3 Some examples of edge-magic labeled graphs

Conjecture 2.2 Every tree is edge-magic.

Some particular families of trees as, for instance, caterpillars have been proven to be edge-magic. However, significant progress in this problem seems to be out of reach at this stage. Therefore, the study of the edge-magicness of particular families of trees may be of interest.

A different family that may be of interest to study is the family of linear forests. Hence, we propose the following open problem.

Problem 2.5 Characterize the set of edge-magic linear forests.

As in the case of trees, some particular families of linear forest have been proven to be edge-magic, as for instance $P_n \cup nK_2$ (n odd), $2P_n$ ($n \in \mathbb{N} \setminus \{1, 2\}$) or, $\cup_{i=1}^n P_i$ among others. However, there are many families of linear forest for which their edge-magic properties are not known, and this is an interesting problem to explore. Another open question that we introduce is the following one.

Problem 2.6 Which graphs of order equal to size are edge-magic?

A special case of EM labelings was introduced by Enomoto et al. [9] in 1998 under the name of super edge-magic (SEM) labelings. Previously, in 1991 Acharya and Hegde introduced the concept of strongly indexable graphs in [1]. It turns out that the sets of super edge-magic graphs and of strongly indexable graphs are the same.

Definition 2.4 ([1]) A (p, q)-graph G is *strongly (k, d)-indexable* if there is a bijective function $f : V(G) \rightarrow [0, p-1]$, such that the set $S = \{f(u) + f(v) : uv \in E(G)\}$ forms an arithmetic progression of $|E|$ terms with first term k and difference d. Then, f is called a *strongly (k, d)-indexable labeling*. For $d = 1$, the labeling is simply called *strongly indexable*.

Figure 2.12b shows a strongly indexable labeling of a caterpillar.

Definition 2.5 ([9]) An edge-magic labeling of a (p, q)-graph G is said to be *super edge-magic* if it has the extra property that $1 \leq f(v) \leq p$. Then G is called a *super edge-magic graph*.

Figure 2.4 shows examples of super edge-magic graphs.

One of the key ideas when dealing with SEM labelings is that to obtain a SEM labeling of a graph it is enough to exhibit the labels of the vertices.

Lemma 2.1 ([10]) *A (p, q)-graph G is super edge-magic if and only if there exists a bijective function $f : V(G) \rightarrow [1, p]$ such that the set $S = \{f(u) + f(v) : uv \in E(G)\}$ consists of q consecutive integers. In such case, f extends to a super edge-magic labeling of G with magic sum $k = p + q + s$, where $s = \min(S)$.*

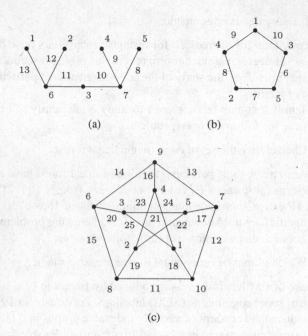

(a)   (b)

(c)

Fig. 2.4 Some examples of super edge-magic labeled graphs

Proof [1]First, assume that such a function f exists and let $xy \in E(G)$ so that $f(x) + f(y) = s$. Then, f extends to the domain $V(G) \cup E(G)$ in the following manner. Let $f(uv) = p+q+s-f(u)-f(v)$ for any edge $uv \in G$. Then, $f(E(G)) = [p+1, p+q]$. Conversely, if G is super edge-magic with a super edge-magic labeling g with magic sum k, then

$$S = \{k - g(uv) : uv \in E(G)\} = [k - (p + q), k - (p + 1)].$$

$\square$

Example 2.1 Figure 2.5 shows (a) a bijective function from the set of vertices of the cycle C_5 to the set $[1, 5]$, whose set $S = \{\bar{4}, \bar{5}, \ldots, \bar{8}\}$ consists of five consecutive integers, and (b) the induced super edge-magic labeling of C_5. The smallest induced sum of the labels of the vertices is replaced by the largest possible label.

Unless otherwise specified, whenever we refer to a function as a super edge-magic labeling we will assume that its domain is the set of vertices. That is, a function f as in Lemma 2.1.

Let $f : V(G) \to [1, p]$ be a super edge-magic labeling of a (p, q)-graph G. The *s-complementary labeling* of f, denoted by $\bar{f}$, is the labeling defined by the rule: $\bar{f}(v) = p + 1 - f(v)$, for all $v \in V(G)$. Clearly, $\bar{f}$ is a super edge-magic labeling of G.

---

[1]With permission from [10], ©2001, Elsevier.

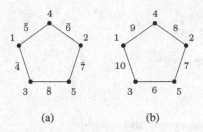

Fig. 2.5 (a) A labeling of the vertices of C_5 and (b) the induced SEM labeling

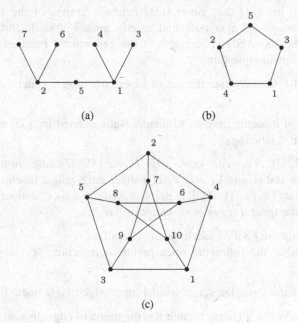

Fig. 2.6 The s-complementary labelings of the super edge-magic labelings of the graphs of Fig. 2.4

Figure 2.6 shows the s-complementary labelings of the graphs shown in Fig. 2.4. According to Lemma 2.1, we only use the vertex labels.

We will devote many lines of this book to the study of super edge-magic labelings, since they play a crucial role in the world of graph labelings. Therefore, the reader will find many interesting open problems on super edge-magic labelings in the forthcoming pages. At this point we believe that it may be interesting to present the following problem next, since it can be a good problem to work on for anyone who wants to get a first taste of problems of this type.

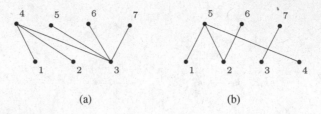

Fig. 2.7 Two special super edge-magic labeled graphs

It was shown in [11] that generalized Petersen graphs of the form $P(n, 2)$ are super edge-magic, if n is odd, and strictly greater than 1. But in general, it is unknown the super edge-magicness of the generalized Petersen graph. This motivates the following question.

Problem 2.7 ([3]) Characterize the set of super edge-magic generalized Petersen graphs.

For the case of bipartite graphs, Muntaner-Batle defined in [25] an additional restriction for SEM labelings.

Definition 2.6 ([25]) A *special super edge-magic (SSEM) labeling* of a bipartite (p, q)-graph with stable sets V_1 and V_2 is a super edge-magic labeling f with the extra property that $f(V_1) = [1, |V_1|]$. A graph that admits a special super edge-magic labeling is called a *special super edge-magic graph*.

Figure 2.7 shows (S)(S)EM labelings of caterpillars.

We present next, the following open problems regarding special super edge-magic graphs.

Problem 2.8 Characterize the set of special super edge-magic linear forests.

A graph is said to be a *galaxy* , when it is the union of edge disjoint stars.

Problem 2.9 Characterize the set of special super edge-magic galaxies.

Another related concept to the one of SEM labelings is the one of (k, d)-arithmetic labelings. It was originally introduced by Acharya and Hegde [1].

Definition 2.7 ([1]) A graph G is (k, d)-*arithmetic* if there is a bijective function $f : V(G) \rightarrow D \subset \mathbb{N}$, such that the set $S = \{f(u) + f(v) : uv \in E(G)\}$ forms an arithmetic progression of $|E|$ terms with first term k and difference d. Then, f is called a (k, d)-*arithmetic labeling* .

An $(8, 2)$-arithmetic labeling of the path P_5 and a $(6, 2)$-arithmetic labeling of C_5 are shown in Fig. 2.8.

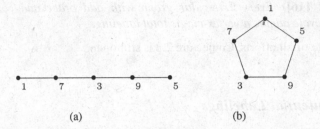

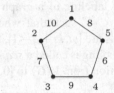

Fig. 2.8 (a) A $(8, 2)$-arithmetic labeling of P_5 and (b) a $(6, 2)$-arithmetic labeling of C_5

Fig. 2.9 A super VMTL of C_5

2.1.3 Vertex-Magic Total Labelings

Another magic total labeling that has also been studied is a kind in which the sum of the labels of all edges adjacent to each vertex v, plus the label of v itself, is constant. This labeling was introduced by MacDougall et al. [23] and was called vertex-magic total labeling.

Definition 2.8 ([23]) A *vertex-magic total labeling (VMTL)* of a (p, q)-graph G is a bijective function $f : V(G) \cup E(G) \rightarrow [1, p+q]$ such that $f(x) + \sum_{y \in N_G(x)} f(xy) = k$, for some $k \in \mathbb{Z}$ and for every $x \in V(G)$. A VMTL is called *strong (super)* if the vertices are labeled with the largest (smallest) available integers. If a graph admits a strong (super) vertex-magic total labeling is called a *strong (super) vertex-magic total graph*.

If we move all labels of Fig. 2.5b clockwise one position, we obtain a strong VMTL of the cycle C_5. Furthermore, Fig 2.9 shows a super VMTL of the cycle C_5.

Problem 2.10 Which families of 2-regular graphs admit a vertex-magic total labeling?

In fact, MacDougal et al. [23] state the following conjecture.

Conjecture 2.3 Every r-regular graph, $r > 1$, with the exception of $2K_3$ has a vertex-magic total labeling.

Some interesting results have appeared in relation to this problem. The following result found in [16] is a good example.

Theorem 2.1 ([16]) *Every 2r-regular graph with odd order and containing a hamiltonian cycle admits a vertex-magic total labeling.*

But in spite of all efforts, Conjecture 2.3 is still open.

2.1.4   Sequential Labelings

Grace introduced in [14] the notion of sequential labeling.

Definition 2.9 ([14]) A *sequential* labeling of a graph G of size q is an injective function $f : V(G) \rightarrow [0, q-1] \subset \mathbb{Z}$ such that when each edge uv is labeled $f(u) + f(v)$, the resulting edge labels are $[m, m+q-1]$, for some positive integer m. A graph that admits a sequential labeling is called a *sequential graph*. For a tree, the function f is allowed to be an injection from $V(G)$ to $[0, q]$.

It is worth to mention that Chang, Hsu, and Rogers had introduced in [7] the concept of *strongly c-harmonious* labeling that turns out to be equivalent to the concept of sequential labeling, except for the fact that for trees, exactly one vertex label is allowed to be used twice.

Figure 2.10 shows some examples of sequential graphs.

The following is an interesting problem about sequential labelings introduced by Ichishima and Oshima in [17].

Problem 2.11 ([17])   Characterize the sequential graphs of the form $mK_{s,t}$.

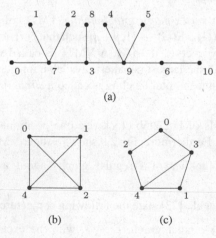

(a)

(b)   (c)

Fig. 2.10   Some examples of sequential graphs

2.1.5 Edge-Antimagic Labelings

Another labeling that we will consider is called (a, d)-edge-antimagic vertex labeling [33], which is equivalent to a strongly (k, d)-indexable labeling [1].

Definition 2.10 ([1, 33]) An (a, d)-*edge-antimagic vertex labeling* [(a, d)-EAV] of a (p, q)-graph G is an injective mapping $f : V(G) \rightarrow [1, p]$ such that the set $\{f(x) + f(y) : xy \in E(G)\}$ is $\{a + id : i \in [0, q - 1]\}$. A graph that admits an (a, d)-edge-antimagic vertex labeling is called an (a, d)-*edge-antimagic vertex graph*.

Figure 2.12e shows a $(3, 2)$-EAV labeling of a caterpillar.

The (a, d)-edge-antimagic total labeling was also introduced in [33] as a variation of Bodendiek and Walther's (a, d)-antimagic labeling [4].

Definition 2.11 ([33]) An (a, d)-*edge-antimagic total* [(a, d)-EAT] labeling of a (p, q)-graph G is a one to one mapping $f : V(G) \cup E(G) \rightarrow [1, p + q]$ such that the set $\{f(u) + f(uv) + f(v) : uv \in E(G)\}$ is an arithmetic progression of q terms starting at a and of difference d. Such a labeling is called *super* if the smallest possible labels appear on the vertices. A graph that admits a (super) edge-antimagic total labeling is called a *(super) edge-antimagic total graph*.

Figure 2.12f shows a $(10, 2)$-EAT labeling of a caterpillar.

2.1.6 Harmonious Labelings

Harmonious labelings were defined by Graham and Sloane [15] as part of their study of additive basis and are applicable to error-correcting codes.

Definition 2.12 ([15]) A (p, q)-graph with $p \leq q$ is called *harmonious* if it is possible to label the vertices with distinct integers (mod q) in such a way that the edge sums are also distinct (mod q). A tree is harmonious if there is a labeling of the vertices in which exactly two vertices have the same label (mod q) and such that the condition on the edge sums holds. Any labeling of this type is called a *harmonious labeling*.

Harmonious labelings have called the attention of many researches since they were introduced in 1980, and they have a place among the most popular labelings in the literature. In fact, if we would have to choose two labelings based on their popularity, our choice would be graceful and harmonious labelings with no doubt. This is the reason why we devote Chap. 4 entirely to this type of labelings.

Figure 2.11 shows examples of harmonious labeled graphs. In Chap. 4, we will encounter the conjecture that states that all trees are harmonious, and this is probably one of the hardest and more popular problems in graph labelings. But it is also true that there are many other open problems that are worth to be studied. Let us see some examples.

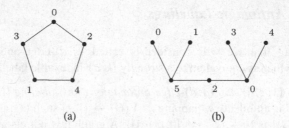

(a)   (b)

Fig. 2.11 Some examples of harmonious labeled graphs

Problem 2.12 Which 2-regular graphs are harmonious?

Problem 2.13 Which unicyclic graphs are harmonious?

Problem 2.14 Which graphs of equal order and size are harmonious?

Related with sequential labelings we find the following problem in [12].

Problem 2.15 ([12]) Does there exist a graph that can be harmoniously labeled but not sequentially labeled?

We feel that the following problem is an interesting one, although it seems that it has not attracted too such attention so far.

Problem 2.16 Characterize the set of harmonious lobsters.

2.1.7   Cordial Labelings

A variation of both, graceful and harmonious labelings, was introduced by Cahit in [6].

Definition 2.13 ([6]) Let f be a function from the vertices of G to the set $\{0, 1\}$ and, for each edge xy assign the label $|f(x) - f(y)|$. If the number of vertices labeled 0 and the number of vertices labeled 1 differ by at most 1, and the number of edges labeled 0 and the number of edges labeled 1 differ at most by 1, then f is called a *cordial labeling* of G and G a *cordial graph*.

Figure 2.12h shows a cordial labeling of a caterpillar.

We find it interesting to provide an example of a graph for which we show different labelings. We do this in Fig. 2.12, where we introduce a caterpillar that has been labeled in different ways as for instance, gracefully, harmoniously, etc.

Exercise 2.1 Find as many types as possible of different labelings (introduced above) for the graphs P_n and C_n, when they exist.

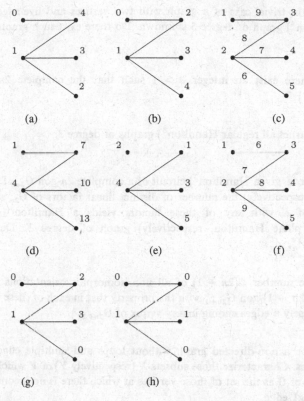

Fig. 2.12 Different labelings of a caterpillar: (**a**) α and β valuation. (**b**) sequential and strongly indexable. (**c**) (S)(S)EM; (**d**) (8, 3)-arithmetic. (**e**) (3, 2)-EAV. (**f**) (10, 2)-EAT. (**g**) harmonious and (**h**) cordial

2.2 Historical Remarks

When trying to identify the starting point of graph labeling, at least as an active area of research in combinatorics, we should refer to the symposium held in Smolenice in June 1963, *Theory of Graphs and its Applications*, where two famous problems were formulated (Fig. 2.13). Ringel [28] conjectured in Problem 25 that every complete graph of order $2n + 1$ can be decomposed into $2n + 1$ subgraphs which are all isomorphic to a given tree with n edges. In Problem 27, Sedláček [32] generalized the number-theoretical notion of magic squares to that of magic graphs as follows: a finite connected graph G without loops or multiple edges is called *magic* if there is a real-valued valuation of the edges of G with the properties that (1) distinct edges have distinct values assigned, and (2) the sum of values assigned to all edges incident to a given vertex x is the same for all vertices v of G. He raised the problem consisting in finding necessary and sufficient conditions for a graph to be magic. Moreover, once we know that a graph is magic does it admit a prime-

19. Except the trivial case of a graph with two vertices and five edges no regular plane Hamilton[3]) graph of degree 5 is known. Do there exist such graphs?

A. KOTZIG

20. Does there exist an integer $n > 1$ such that the complete $2n$-gon is not Hamiltonian?[3])

A. KOTZIG

21. To construct all regular Hamilton[3]) graphs of degree 3.

A. KOTZIG

22. Let K be a given Hamilton[3]) circuit of a complete $2n$-gon G_{2n}. Denote by Λ_n (and Λ_n^*, Λ_n^0, respectively) the number of distinct linear factors of G_{2n} such that the composition of K with any of these factors yields a Hamilton (or bipartite Hamilton or plane Hamilton, respectively) graph of degree 3. Determine the functions Λ_n, Λ_n^*, Λ_n^0.

A. KOTZIG

23. Find the number $\Omega(2n + 1)$ of all non-isomorphic orientations of edges of the complete $(2n + 1)$-gon G_{2n+1} with the property that in each of these orientations there are precisely n edges ending in any vertex of G_{2n+1}.

A. KOTZIG

24. Let G be a non-directed graph without loops and multiple edges, with the set V of vertices. Characterize those subsets X (respectively Y) of V which arise from an orientation of G as the set of those vertices at which there is no incoming (respectively outgoing) edge.

A. KOTZIG

25. It is conjectured that the complete $(2n + 1)$-gon can be decomposed into $2n + 1$ subgraphs which are all isomorphic to a given tree with n edges.

G. RINGEL

26. *Graphs assigned to groups.* Let H be a finite group with elements $a, b, c, \ldots$ and the identity element e. If $a, b \in H$, an element $c \in H$ is uniquely determined so that $abc = e$; then also $bca = cab = e$.

A triple x, y, z of elements from H with $xyz = e$ will be termed an e-triple; two e-triples are equal if one arises from the other by a cyclic permutation. An e-triple (x, y, z) is called regular if $x \neq y \neq z \neq x$, otherwise it is singular. Analogously, a pair $\{x, y\}$ of elements in H is regular if it determines a regular e-triple, otherwise it is singular (especially $\{x, x\}$ is singular for $x \in H$).

---

[3]) Hamiltonian in sense of KOTZIG. See p. 63.

Fig. 2.13   Page 162 in the proceedings of the Symposium held in Smolenice in June 1963

valued valuation, that is, all real numbers assigned to edges are primes? Two years later, in the international symposium of Roma, Rosa introduced the four types of valuations to attack the problem raised by Ringel. Stewart provided in [34, 35] a detailed description of magic graphs.

Although many interesting papers were to come in the next years, it was in 1996, in Kalamazoo, Michigan, when a new impulse was given to the area.

A very well-known meeting all over the world in the area of graph theory and related topics was the *8th Quadrennial International Conference of Graph Theory, Combinatorics, Algorithms and Applications*, 1996, Kalamazoo, MI, USA. Many leading figures in the subject met in the conference, among them Paul Erdős (March 26, 1913, Budapest-September 20, 1996, Varsow) and Gerard Ringel (October 28, 1919, Kollubrum, Austria–June 24, 2008, Santa Cruz, California). Dr Ringel had been working in what appeared to be a new graph labeling problem (although it turned out that this was not the case) together with A. Lladó and O. Serra, that can be explained as follows: for any simple graph, label the vertices and the edges of the graph injectively with consecutive numbers so that the sum of the label of any edge with the labels of the two vertices incident with the edge is always the same. They called such types of labelings edge-magic, and the graphs admitting them also take the same name. In Ringel's talk, that took place in the morning, he was explaining the results that they had obtained during their research. For instance, he showed that all odd cycles are edge-magic and asked what was the case with even cycles, among other things. Paul Erdős was in the audience and it was an unfortunate casualty that during the talk he suffered a heart attack. An ambulance was called and appeared in the spot in no time. It is interesting to notice that while he was taken to the ambulance on the stretcher, he was requesting his briefcase to the nurse, since he claimed that he wanted to keep working. Of course, such a demand was not satisfied by the nurse. Once in the hospital Erdős got an important surgery and he told the doctors that he wanted to assist to the conference's dinner that was taking place that same night. The reason is that the dinner was to his honor, and he did not want to miss such an event. It was about 9:00 p.m. and the dinner was about to start and he appeared, surrounded by doctors and by different medical machines. After having eaten, some mathematicians gave some brief talks and at the end the whole room joined for a big applause dedicated to Paul Erdős. It was a generalized surprise when Erdős got up from his chair and started to give a talk related to the topic he was listening when he got the attack. That is to say, edge-magic labelings. In this talk, he asked several questions that at that moment everyone thought that were open questions, although after some time it was proven that many of the questions had been solved long time ago. Among them: How dense can an edge-magic graph be? Then, he said "At this point, I would be happy if someone can characterize edge-magic complete graphs."

It happened that the second question had already been solved by Kotzig and Rosa long time ago, although they had never published their result in any journal. The only complete edge-magic graphs are K_i, for $i = 1, 2, 3, 5, 6$. The first question was open at that time and was studied and solved later on by Oleg Pikhurko in [26].

Furthermore many of the problems proposed in Ringel's talk had been also solved by Kotzig and Rosa in [18, 21, 22]. For instance, they had showed that C_n is edge-magic for all $n \in \mathbb{N} \setminus \{1, 2\}$. Due to the fact that their papers were hard to find, since it has been already mentioned that they had not been published in formal journals, it follows that many researchers were not aware of this, and hence they worked on these problems. In particular, several people found the edge-magicness of C_n. For instance, Ichishima (personal communication), Craft and Tesar [8] and Roddity et al. [30]. Craft and Tesar had also characterized edge-magic complete graphs in the same paper. Similarly, Lladó (personal communication) had found a solution of this problem. We mention that the original proof by Kotzig and Rosa is quite lengthy, since it relies on results with very lengthy proofs. On the other hand, both proofs by Craft and Tesar and by Lladó were quite shorter. In summary, both proofs follow the idea that for $n > 6$, the graph K_n contains too many edges to admit an edge-magic labeling. The graph K_4 is not edge-magic due to parity reasons. Another researcher [37] gave a very elegant combinatorial argument to attack this problem. Unfortunately his argument is not strong enough to completely characterize all edge-magic complete graphs.

In spite of all this, we have the strong conviction that this chain of events motivated many people, to work on the topic of graph labelings, and thanks to this, this area has experimented a fast development during the last two decades. Nevertheless, and in spite of all the people working in the subject, there are many open and very interesting and challenging problems to work on. We are sure that the reader will find many of them if she/he continues her/his journey through the book.

References

1. Acharya, B.D., Hegde, S.M.: Arithmetic graphs. J. Graph Theory **14**, 257–299 (1990)
2. Bača, M., Miller, M.: Super Edge-Antimagic Graphs. BrownWalker Press, Boca Raton (2008)
3. Bača, M., Baskoro, E.T., Simanjuntak, R., Sugeng, K.A.: Super edge-antimagic labelings of the generalized Petersen graph $P(n, (n-1)/2)$. Utilitas Math. **70**, 119–127 (2006)
4. Bodendiek, R., Walther, G.: Arithmetish a timagishe graphen. In: Graphentheorie III. Bi-Weiss Ver., Mannheim (1993)
5. Brankovic, L., Murch, C., Pond, J., Rosa, A.: Alpha-size of trees with maxi- mum degree three and perfect matching. In: Proceedings of AWOCA 2005, pp. 47–56 (2005)
6. Cahit, I.: Cordial graphs: a weaker version of graceful and harmonious graphs. Ars Comb. **23**, 201–207 (1987)
7. Chang, G.J., Hsu, D.F., Rogers, D.G.: Additive variations on a graceful theme: some results on harmonious and other related graphs. Congr. Numer. **32**, 181–197 (1981)
8. Craft, D., Tesar, E.H.: On a question by Erdős about edge-magic graphs. Discrete Math. **207**(1–3), 271–276 (1999)
9. Enomoto, H., Lladó, A., Nakamigawa, T., Ringel, G.: Super edge-magic graphs. SUT J. Math. **34**, 105–109 (1998)
10. Figueroa-Centeno, R.M., Ichishima, R., Muntaner-Batle, F.A.: The place of super edge-magic labelings among other classes of labelings. Discrete Math. **231**(1–3), 153–168 (2001). http://dx.doi.org/10.1016/S0012-365X(00)00314-9

11. Fukuchi, Y.: Edge-magic labelings of generalized Petersen graphs $p(n, 2)$. Ars Comb. **59**, 253–257 (2001)
12. Gallian, J.A.: A dynamic survey of graph labeling. Electron. J. Comb. **19**(DS6) (2016)
13. Golomb, S.W.: How to number a graph. In: Graph Theory and Computing, pp. 23–37. Academic, New York (1972)
14. Grace, T.: On sequencial labelings of graphs. J. Graph Theory **7Ish**, 195–201 (1983)
15. Graham, R.L., Sloane, N.J.A.: On additive bases and harmonious graphs. SIAM J. Algebr. Discrete Methods **1**, 382–404 (1980)
16. Gray, I.D., MacDougall, J.A.: Vertex-magic labelings of regular graphs II. Discrete Math. **309**(20), 5986–5999 (2009)
17. Ichishima, R., Oshima, A.: On the super edge-magic deficiency and α-valuations of graphs. Ars Comb. **129**, 157–163 (2016)
18. Kotzig, A.: On a class of graphs without magic valuations. Reports of the CRM **CRM-136** (1971)
19. Kotzig, A.: β-valuations of quadratic graphs with isomorphic components. Utilitas Math. **7**, 263–279 (1975)
20. Kotzig, A.: Recent results and open problems in graceful graphs. Congr. Numer. **44**, 197–219 (1984)
21. Kotzig, A., Rosa, A.: Magic valuations of finite graphs. Can. Math. Bull. **13**, 451–461 (1970)
22. Kotzig, A., Rosa, A.: Magic valuations of complete graphs. Publ. CRM **175** (1972)
23. MacDougall, J.A., Miller, M., Slamin, Wallis, W.D.: Vertex-magic total labelings of graphs. Utilitas Math. **61**, 3–21 (2002)
24. Marr, A.M., Wallis, W.D.: Magic Graphs, 2nd edn. Birkhaüser, New York (2013)
25. Muntaner-Batle, F.A.: Special super edge-magic labelings of bipartite graphs. J. Comb. Math. Comb. Comput. **39**, 107–120 (2001)
26. Pikhurko, O.: Dense magic graphs and thin additive bases. Discrete Math. **306**(17), 2097–2107 (2006)
27. Redd, T.: Graceful graphs and graceful labelings: two mathematical formulations and some other new results. Congr. Numer. **164**, 17–31 (2003)
28. Ringel, G.: Problem 25. In: Theory of Graphs and Its Application (Proc. Sympos. Smolenice 1963), p. 162. Nakl. CSAV, Praha (1964)
29. Ringel, G., Lladó, A.: Another tree conjecture. Bull. Inst. Comb. Appl. **18**, 83–85 (1996)
30. Roddity, Y., Bachar, T.: A note on edge-magic cycles. Bull. Inst. Comb. Appl. **29**, 94–96 (2000)
31. Rosa, A.: On certain valuations of the vertices of a graph. In: Theory of Graphs (Internat. Symposium, Rome, July 1966), pp. 349–355. Gordon and Breach/Dunod, New York/Paris (1967)
32. Sedláček, J.: Problem 27. In: Theory of Graphs and its Application (Proc. Sympos. Smolenice 1963), pp. 163–164. Nakl. CSAV, Praha (1964)
33. Simanjuntak, R., Bertault, F., Miller, M.: Two new (a, d)-antimagic graph labelings. In: Proc. of Eleventh Australasian Workshop on Combinatorial Algorithms, pp. 149–158 (2000)
34. Stewart, B.M.: Magic graphs. Can. J. Math. **18**, 1031–1059 (1966)
35. Stewart, B.M.: Supermagic complete graphs. Can. J. Math. **19**, 427–438 (1967)
36. Truszczyński, M.: Graceful unicyclic graphs. Demonstatio Math. **17**, 377–387 (1984)
37. Valdés, L.: Edge-magic K_p. In: Thirty-Second South-Eastern International Conference on Combinatorics, Graph Theory and Computing, Baton Rouge, LA, vol. 153, pp. 107–111 (2001)

Chapter 3
Super Edge Magic Labelings: First Type of Relations

3.1 Basic Definitions

The number of different types of graph labelings has become enormous during the last five decades. A good proof of that is the survey by Gallian [15]. It seems that researchers consider each labeling separately from the rest of labelings. The title of the second survey paper on graph labelings published by Gallian [14], "A guide to the graph labeling zoo" reflects very well this fact. Therefore, it makes sense to ask whether there exists a link among labelings, or at least among some of the most well-known labelings. The goal of this chapter is to show that there are indeed many bridges relating different labelings. Although we believe that super edge-magic labelings are the center of the graph labeling network [10, 13], some other interesting relations have appeared recently in [2, 5, 6]. To emphasize these relations is one of the key points of this book and this is the reason why we dedicate this chapter to them. Furthermore, in Chap. 6, we shall develop new tools that will allow us to find more relations with new graphs and labelings.

The first relation established is with sequential labelings. For integers $m \leq n$, we denote the set $\{m, m+1, ..., n\}$ by $[m, n]$.

Theorem 3.1 ([10]) *Let G be a (p, q)-graph that is either a tree or has $q \geq p$. If G is super edge-magic, then G is sequential.*

Proof Let f be a super edge-magic labeling of G. Then $\{f(u)+f(v) : uv \in E(G)\} = [m, m+q-1]$. Now, define the function $g : V(G) \to [0, p-1]$ to be the injective function such that $g(v) = f(v) - 1$, for all $v \in V(G)$. Thus, $\{g(u) + g(v) : uv \in E(G)\} = [m-2, m+q-3]$. Hence, g is a sequential labeling of G. $\square$

Figure 2.12b, c may serve as an example of how the functions f and g described in the above proof are related. A generalization of the previous result is stated next.

Proposition 3.1 ([13]) *Let G be a (p, q)-SEM graph. Then for every positive integer d, there exists a positive integer k such that G is (k, d)-arithmetic.*

© The Author(s) 2017   33
S.C. López, F.A. Muntaner-Batle, *Graceful, Harmonious and Magic Type Labelings*,
SpringerBriefs in Mathematics, DOI 10.1007/978-3-319-52657-7_3

Proof Assume that the vertices of G are named after the labels of some SEM labeling. For a fix d we consider the labeling g of G with $g(i) = 1 + (i - 1)d$, for $i \in [1, p]$. Since $\{i + j : ij \in E(G)\}$ are consecutive numbers, the numbers $\{g(i) + g(j) : ij \in E(G)\} = \{2 + (i + j - 2)d : ij \in E(G)\}$ form an arithmetic progression with difference d.   □

Figure 2.12c, d show how we can get an $(8, 3)$-arithmetic labeling of a tree from a super edge-magic labeling of the same tree.

The following lemma found in [10] shows a relationship existing between super edge-magic graphs and harmonious graphs.

Lemma 3.1 ([10]) *If G is a super edge-magic (p, q)-graph, then G is harmonious whenever $q \geq p$ or G is a tree.*

Proof First of all, we assume that G is a tree. If f is any super edge-magic labeling of G, then it suffices to reduce all labels modulo $p-1$ and we are done. By Theorem 3.1 G is sequential. Thus, if G is not a tree, then G admits an injective labeling $g : V(G) \to D \subset [0, q]$ such that $\{g(a) + g(b) : ab \in E(G)\} = [m, m + q - 1]$, for some $m \in \mathbb{N}$. Hence, all equivalence classes of the group $\mathbb{Z}_q$ have exactly one representative in $[m, m + q - 1]$, and therefore g is also a harmonious labeling of G.
                                                                                                  □

Figure 2.12c, g perfectly illustrate Lemma 3.1.

The following results show relations among α-valuations and special super edge-magic labelings.

Theorem 3.2 ([10]) *If G is a special super edge-magic bipartite $(p, p - 1)$-graph, then G has an α-valuation.*

Proof Let V_1 and V_2 be the stable sets of G. Assume that f is a special super edge-magic labeling of G with $f(V_1) = [1, p_1]$ and $f(V_2) = [p_1 + 1, p_1 + p_2]$. Let $g : V(G) \to [0, p - 1]$ be the labeling defined by the rule:

$$g(v) = \begin{cases} f(v) - 1, & \text{if } v \in V_1, \\ 2p_1 + p_2 - f(v), & \text{if } v \in V_2. \end{cases}$$

Next, we prove that g is an α-valuation of G with characteristic $p_1 - 1$. First observe that $g(V_1) = [0, p_1 - 1]$ and $g(V_2) = [p_1, p - 1]$. Also, if $uv \in E(G)$ with $u \in V_2$ and $v \in V_1$, we have $|g(u) - g(v)| = g(u) - g(v) = 2p_1 + p_2 - f(u) - (f(v) - 1) = 2p_1 + p_2 + 1 - (f(u) + f(v))$.

Since $\{f(u) + f(v) : uv \in E(G)\}$ is a set of consecutive natural numbers and $\max\{f(u) + f(v) : uv \in E(G)\} \leq 2p_1 + p_2$, it follows that g is an α-valuation of G.   □

As an example, observe that Fig. 2.12a shows the α-valuation of a caterpillar obtained from the special super edge-magic labeling of the same caterpillar shown in Fig. 2.12c.

Rosa showed that all graphs that admit α-valuations are bipartite.

Lemma 3.2 ([21]) *Let G be a (p, q)-graph that admits an α-valuation. Then G is bipartite.*

Proof Assume to the contrary that G admits an α-valuation, namely f, but G is not bipartite. Then G contains a cycle of odd order C_r as a subgraph. Let $V(C_r) = \{v_i\}_{i=1}^r$ and $E(C_r) = \{v_iv_{i+1}\}_{i=1}^{r-1} \cup \{v_rv_1\}$. Since f is an α-valuation of G and $C_r \subseteq G$, it follows that there exists $k \in [0, q-1]$ such that, for every $v_iv_j \in E(C_r)$ either $f(v_i) \leq k < f(v_j)$ or $f(v_j) \leq k < f(v_i)$. Without loss of generality, assume that $f(v_1) > k \geq f(v_2)$. Thus, $f(v_2) \leq k < f(v_3), f(v_3) > k \geq f(v_4), \ldots$, and hence $f(v_r) > k \geq f(v_1)$. Therefore, a contradiction has been reached. $\square$

Hence, the converse of Theorem 3.2 follows.

Theorem 3.3 ([10]) *Let G be a $(p, p-1)$-graph with an α-valuation. Then G is special super edge-magic.*

Proof We may assume that G is connected, otherwise we just apply the same reasoning to each connected component of G. Let g be an α-valuation of G. By Lemma 3.2, G is bipartite. Let V_1, V_2 be the stable sets of $V(G)$. We show that if $u \in V_1$, $v \in V_2$, and k is the characteristic of g, then either $g(u) > k$ and $g(v) \leq k$; or $g(u) \leq k$ and $g(v) > k$. We proceed by contradiction. Without loss of generality, assume to the contrary that there exist $u \in V_1$ and $v \in V_2$ such that $g(u) > g(v) > k$. Now, $u \subset V_1$ and $v \in V_2$. Thus, any path joining u and v has even order. Let $P : u = x_1, x_2, \ldots, x_{2l-1}, x_{2l} = v$ be one of such paths. Then, $g(u) = g(x_1) > k$, $g(x_2) \leq k$, $g(x_3) > k, \ldots, g(x_{2l-1}) > k$ and $g(x_{2l}) = g(v) \leq k$. Therefore, a contradiction has been reached.

Hence, without loss of generality, we may assume that $g(V_1) = [0, p_1 - 1]$ and $g(V_2) = [p_1, p-1]$. Since $|E(G)| = p-1$, it follows that the set $\{g(u) - g(v) : uv \in E(G), u \in V_2$ and $v \in V_1\}$ is a set of $p-1$ consecutive integers.

Finally, we show that the function $f : V(G) \to [1, p]$ defined by the rule

$$f(v) = \begin{cases} g(v) + 1, & \text{if } v \in V_1, \\ 2p_1 + p_2 - g(v), & \text{if } v \in V_2, \end{cases}$$

is a special super edge-magic labeling of G. First of all notice that $f(V_1) = [1, p_1]$ and $f(V_2) = [p_1 + 1, p_1 + p_2]$. Now, for each $uv \in E(G)$ with $u \in V_1$ and $v \in V_2$, we obtain $f(u) + f(v) = g(u) + 1 + 2p_1 + p_2 - g(v) = g(u) - g(v) + p_1 + p + 1$. Since $\{g(u) - g(v) : uv \in E(G), u \in V_2$ and $v \in V_1\}$ is a set of $|E(G)|$ consecutive integers, we are done. $\square$

The special super edge-magic shown in Fig. 2.12c can be obtained using the α-valuation shown in Fig. 2.12a. In particular, we obtain the next corollary.

Corollary 3.1 ([10]) *A tree T admits an α-valuation if and only if T is special super edge-magic.*

Exercise 3.1 Prove or disprove the following statement: "all trees admit an α-valuation."

Remark 3.1 Assume that T is a tree that admits an α-valuation f and let A and B be its stable sets. Assume that $|A| \geq |B|$. Without loss of restriction, we can also assume that the characteristic of f is assigned to $|A|$. Otherwise, we can consider the graceful labeling $\bar{f}$ obtained from f and defined by $\bar{f}(v) = |E(T)| - f(v)$ which verifies this condition.

The next results establish a relationship between α-valuations and $(3,2)$-EAV labelings of trees.

Lemma 3.3 ([2]) *Let T be tree that admits an α-valuation with stable sets A and B. If $||A| - |B|| \leq 1$, then T has a $(3,2)$-EAV labeling.*

Proof Let f be an α-valuation of a tree T of order p with characteristic k. By Remark 3.1, we can assume that $|A| \geq |B|$ and that the vertex labeled k belongs to A. Consider the following labeling of the vertices of T.

$$g(v) = \begin{cases} 2f(v) + 1, & \text{if } v \in A, \\ 2(p - f(v)), & \text{if } v \in B. \end{cases}$$

We claim that g is a $(3,2)$-EAV labeling of T. In fact, notice that g is an injective function that assigns the labels $\{1, 3, \ldots, 2k+1\} \cup \{2, 4, \ldots, 2(p-k-1)\}$ to the vertices of T. Since $|A|-|B| \leq 1, k = \lceil p/2 \rceil - 1$ and this union is $[1, p]$. Furthermore, $\{g(u) + g(v) : uv \in E(T)\} = \{2p + 1 - 2(f(v) - f(u)) : uv \in E(T)\}$. Since f is an α-valuation, $\{f(v) - f(u) : uv \in E(T)\} = [1, p-1]$, we have $\{g(u) + g(v) : uv \in E(T)\} = \{3, 5, \ldots, 2p-1\}$. Thus, the function g is a $(3,2)$-EAV labeling of T. □

Using the α-labeling found in Fig. 2.12 and Remark 3.1, we get the α-labeling with characteristic 2 shown in Fig. 3.1a. Moreover, using the function described in the proof of the previous lemma, we get the $(3,2)$-EAV labeling shown in Fig. 3.1b.

The next exercise contains the generalization of Lemma 3.3 that Bača et al. obtained for general bipartite graphs with α-valuations.

Exercise 3.2 Prove the following statement: "let G be a $(p, p-1)$-graph with an α-valuation and let $\{V_1, V_2\}$ be the stable sets of G. If $||V_1| - |V_2|| \leq 1$ then mG is $(m+2, 2)$-EAV" [6].

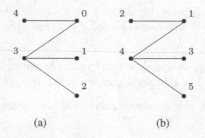

(a)   (b)

Fig. 3.1 Different labelings of a caterpillar: (**a**) An α-valuation with characteristic 2. (**b**) A $(3,2)$-EAV

The next results allow us to characterize trees that admit $(3,2)$-EAV labelings. We leave the proof of the first one as an exercise.

Exercise 3.3 Let T be a tree with stable sets A and B. Prove that if $||A| - |B|| > 1$ then there is no $(3,2)$-EAV labeling of T [2].

Lemma 3.4 ([2]) *Let T be a tree. If T admits a $(3,2)$-EAV labeling, then T admits an α-valuation.*

Proof Let p be the order of T, A and B its stable sets and assume that $1 \in A$. Suppose that g is a $(3,2)$-EAV labeling of T. Thus, if $v \in A$, then $g(v) \in \{1,3,\ldots,p\}$ when p is odd or $g(v) \in \{1,3,\ldots,p-1\}$ when p is even, and if $v \in B$, $g(v) \in \{2,4,\ldots,p-1\}$ when p is odd or $g(v) \in \{2,4,\ldots,p\}$ when p is even.

Consider the labeling $f : V(T) \to [0, p-1]$ defined by

$$f(v) = \begin{cases} (g(v) - 1)/2, & \text{if } v \in A, \\ (2p - g(v))/2, & \text{if } v \in B. \end{cases}$$

Thus, f assigns to the vertices of A the labels $[0, \lceil p/2 \rceil - 1]$ and to the vertices of B the labels $[\lceil p/2 \rceil, p-1]$.

Let $x, y \in V(T)$ such that $x \in A$ and $y \in B$, thus $f(y) - f(x) = (2p + 1 - (g(x) + g(y))/2$. Since $\{g(x) + g(y) : xy \in E(T)\} = \{3, 5, \ldots, 2p - 1\}$, we have $\{f(y) - f(x) : xy \in E(T)\} = [1, p - 1]$, which implies that f is a graceful labeling of T. Since the labels of the vertices in A are less than the labels of the vertices in B, f is an α-valuation of T. This proves the result. $\square$

Figure 3.1 perfectly illustrates the proof of Lemma 3.4. Using the two previous results the following theorem can be proved.

Theorem 3.4 ([2]) *A tree admits a $(3,2)$-EAV labeling if and only if T admits an α-valuation and $||A| - |B|| \leq 1$, where A and B are the stable sets of T.*

Another interesting relation involving cordial graphs and super edge-magic graphs is stated next.

Exercise 3.4 Prove that if a graph G is super edge-magic, then G is cordial [10].

A summary of the relations established above involving SEM labelings can be visualized in Fig. 3.2. We conclude this section by presenting some relations involving (a,d)-EAT labelings.

Theorem 3.5 ([3]) *Let G be a (p,q)-graph which admits a total labeling and whose edge labels constitute an arithmetic progression with difference d. Then the following are equivalent.*

 (i) G has an EM labeling with magic constant k.
(ii) G has a $(k - (q - 1)d, 2d)$-EAT labeling.

(1) If tree or $p \le q$.

(2) If tree or $q \ge p$.

(3) If $q = 2p - 3$.

(4) If tree.

(5) If $(p, p - 1)$ bipartite graph.

(6) If tree.

(7) If tree, and
$$\|A| - |B\| \le 1.$$

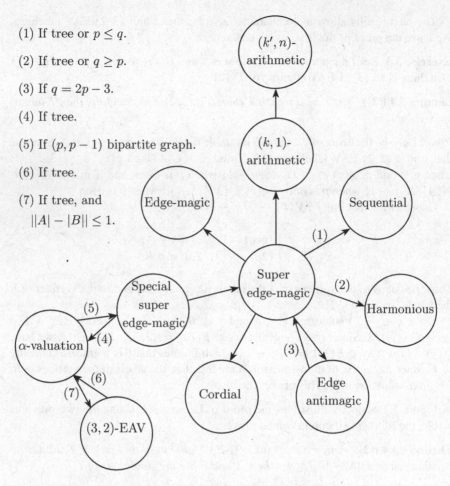

Fig. 3.2 Relations for the existence of labelings for a (p, q)-graph G

Proof Let f be an edge-magic total labeling for G, with $f(E) = \{b, b + d, \ldots, b + (q-1)d\}$ for some $b \in [1, p + q - (q-1)d]$. Define e_i as the edge labeled $b + id$, $i \in [0, q - 1]$.

Next define a new labeling f'

$$f'(x) = f(x) \qquad \text{for } x \in V,$$
$$f'(e_i) = 2b + (q - 1)d - f(e_i) \text{ for } i \in [0, e - 1].$$

Under the labeling f' the set of edge-weights of edges in G is equal to

$$\{k + 2b + (q - 1)d - 2f(e_i) : i \in [0, q - 1]\}$$
$$= \{k + 2b + (q - 1)d - 2(b + id) : i \in [0, q - 1]\}$$

$$= \{k + (q-1)d - i(2d) : i \in [0, q-1]\}$$
$$= \{k - (q-1)d + i(2d) : i \in [0, q-1]\}.$$

The converse can be proved in a similar way. $\square$

The labelings of Fig. 2.12c, f have been obtained from each other following the same idea found in the above proof.

The following exercise illustrates the relation existing between edge-magic labelings with magic constant k and $(k - (q-1)d, 2d)$-EAT labelings.

Exercise 3.5 Let G be a $(p, p-1)$-graph, with even $p \geq 4$. Prove that if G admits an α-valuation, then mG admits a super $(b, 1)$-EAT labeling for every $m \geq 1$ [6].

Exercise 3.6 Prove that if a (p, q)-graph G has an (a, d)-edge-antimagic vertex labeling then

 (i) G has an $(a + p + 1, d + 1)$-edge-antimagic total labeling,
(ii) G has an $(a + p + q, d - 1)$-edge-antimagic total labeling [6].

3.2 Super Edge-Magic Labelings

As we have seen in Sect. 3.1, super edge-magic labelings play an important role when studying relations among labelings. In this section, we will recover some of the main properties of such labelings and, later on, in Sect. 6.3, we will see how a product of digraphs, namely the $\otimes_h$-product, can be used to obtain new families of super edge-magic graphs from existing ones.

Recall that, an *edge-magic labeling* (EM) of a (p, q)-graph G is a bijection $f :$ $V(G) \cup E(G) \rightarrow [1, p+q]$ such that $f(u) + f(uv) + f(v) = k$, for all $uv \in E(G)$. If it has the extra property that $1 \leq f(v) \leq p$, then f is said to be a *super edge-magic labeling* (SEM) of G. The constant k is called the *valence*, the *magic weight* or *the magic sum* of the labeling f. A further restriction of SEM labelings was introduced by Muntaner-Batle in [20], which only makes sense for bipartite graphs. A *special super edge-magic (SSEM) labeling* of a bipartite (p, q)-graph with stable sets V_1 and V_2 is a super edge-magic labeling f with the extra property that $f(V_1) = [1, |V_1|]$.

By Lemma 2.1, to obtain a SEM labeling of a graph is enough to exhibit the labels of the vertices. From Lemma 2.1, we also obtain the following results.

Corollary 3.2 ([10]) *Let G be a super edge-magic (p, q)-graph and let f be a super edge-magic labeling of G. Then $\sum_{v \in V(G)} f(v)d(v) = qs + q(q-1)/2$, where s is defined as in Lemma 2.1. In particular,*

$$2 \sum_{v \in V(G)} f(v)d(v) \equiv 0 \pmod{q}.$$

Certain graphs whose components are Eulerian are excluded from the class of super edge-magic graphs.

Corollary 3.3 ([10]) *Let G be a (p, q)-graph, where every vertex of G has even degree and $q \equiv 2 \pmod 4$, then G is not super edge-magic.*

In fact, Corollary 3.3 can also be obtained from the following result.

Proposition 3.2 ([10]) *Let G be any super edge-magic graph of size q and let f be a super edge-magic labeling of G. Then, there are exactly either $\lfloor q/2 \rfloor$ or $\lceil q/2 \rceil$ edges between V_e and V_o, where $V_e = \{v \in V(G) : f(v) \text{ is even}\}$ and $V_o = \{v \in V(G) : f(v) \text{ is odd}\}$.*

Exercise 3.7 Prove Proposition 3.2 [10].

Due to Proposition 3.2, we know that the vertex set of any super edge-magic graph must admit a partition into two sets such that the number of edges joining the vertices of the two sets that form the partition is either the ceiling or the floor of half the size of the graph. However, although this property is theoretically interesting, in general it is not easy to check its existence. In fact, in [8] it was shown that the problem of determining the existence of such partition is an NP-complete problem.

The next result states a necessary condition for a regular graph to be super edge-magic.

Corollary 3.4 ([10]) *If G is an r-regular super edge-magic (p, q)-graph, where $r > 0$, then q is odd and the magic sum of any super edge-magic labeling of G is $(4p + q + 3)/2$.*

Proof It is clear that the magic sum of any super edge-magic labeling of an r-regular (p, q)-graph G is:

$$\frac{1}{q} \left(r \sum_{i=1}^{p} i + \sum_{i=p+1}^{p+q} i \right) = \frac{1}{q} \left((p+1)\frac{rp}{2} + \frac{q(2p+q+1)}{2} \right)$$

$$= \frac{1}{q} \left((p+1)q + \frac{q(2p+q+1)}{2} \right) = \frac{4p+q+3}{2}.$$

Since the magic sum is an integer, it follows that q is odd.   □

Super edge-magic labelings of 2-regular graphs have strong connections with super vertex-magic total labelings. According to the previous corollary, the magic sum of these graphs is equal to $(5p + 3)/2$, where p is the order.

Next, we mention a useful result that was found by Enomoto et al. [9].

Lemma 3.5 ([9]) *If a (p, q)-graph is super edge-magic, then $q \leq 2p - 3$.*

Proof By Lemma 2.1 if a (p, q)-graph G is super edge-magic, then there is a bijective function $f : V(G) \to [1, p]$ such that the set $S = \{f(x) + f(y) : uv \in E\}$ is a set of q consecutive integers. However, $\min(S) \geq 1 + 2 = 3$ and $\max(S) \leq (p - 1) + p = 2p - 1$. Hence, $|S| \leq (2p - 1) - 3 + 1 = 2p - 3$. Therefore, $q \leq 2p - 3$.   □

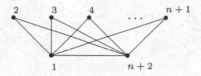

Fig. 3.3 Super edge-magic graphs with $q = 2p - 3$

It is easy to check that the bound given in Lemma 3.5 is tight. For instance, consider the family Δ of super edge-magic labeled graphs that appears in Fig. 3.3.

Notice that all graphs in Δ contain triangles (that is to say, cycles of length 3). In fact, this is not a coincidence. See the following result found in [12].

Theorem 3.6 ([12]) *Let G be a super edge-magic (p,q)-graph with $q \geq 2p - 4$. Then G contains triangles.*

Proof Assume to the contrary that G contains no triangles. Let $f : V(G) \to [1,p]$ be a super edge-magic labeling of G, and let $V(G) = \{v_i\}_{i=1}^{p}$ be such that $f(v_i) = i$ for all $i \in [1,p]$. First observe that since $q \geq 2p - 4$ it follows that either $v_1 v_2 \in E(G)$, or $v_{p-1} v_p \in E(G)$, since the numbers 3 and $2p - 1$ can be expressed uniquely as the sums of two distinct integers in the range from 1 up to p. Assume without loss of generality that $v_1 v_2 \in E(G)$. (Otherwise, it means that $v_{p-1} v_p \in E(G)$, and in the s-complementary labeling $\bar{f}$ of f, which is also super edge-magic, we have $\bar{f}(v_p) = 1$ and $\bar{f}(v_{p-1}) = 2$.) Thus $v_1 v_3 \in E(G)$ since 4 can be expressed uniquely as $1 + 3$ with two distinct integers in the set $[1,p]$. Hence $v_2 v_3 \notin E(G)$ since G contains no triangles, and thus, $v_1 v_4 \in E(G)$. Continuing avoiding triangles in this manner, we conclude that if $d = d_G(v_1)$ then $\{v_1 v_i : i \in [2, d+1]\}$ is a subset of $E(G)$, and none of the vertices $v_2, v_3, \ldots, v_{d+1}$ is adjacent to any another vertex. We have thus accounted for the sums 3 up to $d + 2$. If $d = p - 1$, we are done, since there is no way to obtain the sum $d + 3$ avoiding triangles. Otherwise, if $d < p - 1$ then with the remaining options, the smallest possible sum that can be obtained is $d + 4$, and we have no way to obtain the sum $d + 3$. Therefore, f is not a super edge-magic labeling of G, a contradiction. $\square$

Hence, we have the following corollary.

Corollary 3.5 ([12]) *Let G be a super edge-magic graph of order $p \geq 4$ and size q that contains no subgraph isomorphic to C_3. Then*

$$q \leq 2p - 5.$$

This bound is sharp, since it is not hard to find bipartite graphs which are super edge-magic and reach the bound. See the examples in Fig. 3.4. Thus, the following question rises naturally.

Question 3.1 ([17]) What is the maximum size q of a super edge-magic graph G of order p and girth γ?

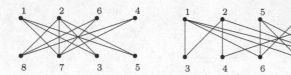

Fig. 3.4 Super edge-magic graphs with $q = 2p - 5$

Next, we discuss what it is known so far about this question. For girth $\gamma = 3$, we have $q \le 2p - 3$ and this bound is tight. For girth $\gamma = 4$, we have $q \le 2p - 5$ and this bound is tight. Let us now study this problem for girth $\gamma = 5$. The following result was established in [17].

Theorem 3.7 ([17]) *There exists an infinite family of super edge-magic graphs of order p, size $q = 2p - 5$ and girth 5.*

Proof Consider the family $\Omega = \{\omega_n : n \in \mathbb{N}\}$ of graphs where each ω_n has order $5n$ and size $10n - 5$. Next we describe the graphs of this family. Let ω_1 be the cycle of order 5. For $n > 1$, let $[0, 5n - 1]$ be the vertex set of ω_n. The graph ω_n consists of n cycles, each one of them called the level L_k, where $k \in [1, n]$. The vertices of the level L_k are $V(L_k) = [5k - 5, 5k - 1]$. Each vertex in L_k is adjacent with exactly one vertex of the level L_{k-1} and with exactly one vertex of the level L_{k+1}, for each $k \in [2, n - 1]$. Consequently the vertices of $L_2, \ldots, L_{n-1}$ have all degree 4 and the vertices of L_1 and L_n have degree 3. Next we describe these adjacencies.

Let $a, b \in V(L_k)$, where $k \in [1, n]$. We denote by $\bar{a}$ and $\bar{b}$ the remainders of a and b modulo 5, respectively. Then $ab \in E(\omega_n)$ if and only if either $\bar{a} = \bar{b} +_5 2$ or $\bar{b} = \bar{a} +_5 2$, where $+_5$ denotes the sum in $\mathbb{Z}_5$. Next, let $a \in V(L_k)$ and $b \in V(L_{k+1})$, $k \in [1, n-1]$. Then $ab \in E(\omega_n)$ if and only if $\bar{b} = \pi(\bar{a})$ when k is odd or $\bar{b} = \pi^{-1}(\bar{a})$ when k is even, where π is the following permutation of the elements of $\mathbb{Z}_5$ in cycle notation: $(0, 4, 1, 2)(3)$.

Figure 3.5 shows the graph ω_3. Note that ω_1 is the cycle C_5 and that ω_2 is the Petersen graph.

Next, we show that the girth of ω_n is 5 for all $n \in \mathbb{N}$. For $n = 1$, $\omega_1 = C_5$, and hence the result is clear. Now, for $n = 2$, we observe that the subgraphs ω_n induced by two consecutive levels are all isomorphic to the Petersen graph, and the girth of the Petersen graph is 5. Hence, any cycle in ω_n of order strictly less than 5 must contain vertices of at least three distinct levels, we can only construct cycles of order at least 6. Therefore, the girth of ω_n is 5, for all $\omega_n \in \Omega$. Finally, we must show that if $\omega_n \in \Omega$, then ω_n is super edge-magic. Let $f : [0, 5n-1] \longrightarrow [1, 5n]$ defined by the rule $f(i) = i+1$. Then $\{f(a)+f(b) : ab \in E(L_k)$ and $k \in [1, n]\} = [10k-6, 10k-2]$ and if $k \in [1, n - 1]$, we have $\{f(a) + f(b) : ab \in E(\omega_n)\ a \in V(L_k)$ and $b \in V(L_{k+1})\} = [10k - 1, 10k + 3]$. Thus, $\{f(a) + f(b) : ab \in E(\omega_n)\} = [4, 10n - 2]$ and $|E(\omega_n)| = |[4, 10n - 2]|$. That is, f is a super edge-magic labeling of ω_n.   □

Up to this point, no more is known about Question 3.1, and we believe that this opens a very interesting research line to be followed.

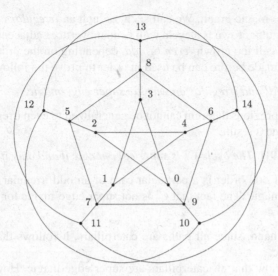

Fig. 3.5   The graph ω_3

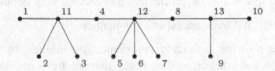

Fig. 3.6   A super edge-magic labeling of a caterpillar

3.2.1   Examples of Edge-Magic and Super Edge-Magic Graphs

This section is devoted to introducing other families of edge-magic and super edge-magic graphs. We will mainly concentrate on SEM graphs, since SEM labelings in many cases imply the existence of many other labelings, as we have seen in Sect. 3.1. We begin with the following result that shows that super edge-magic graphs are not too many when compared with the set of all graphs. We just state the result since we are not ready yet to provide a proof. We will prove the result right after the proof of Theorem 4.10 , since we need this result for our proof.

Theorem 3.8 ([10])   *Almost all graphs are not super edge-magic.*

Thus, we have that super edge-magic graphs are rare among the set of graphs. However, some graphs with certain structure are super edge-magic.

Theorem 3.9 ([9, 18, 19])   *All caterpillars are super edge-magic.*

Exercise 3.8   Prove Theorem 3.9. (Hint. See Fig. 3.6.)

Figure 3.6 shows an example of a caterpillar labeled in a super edge-magic way. Notice that the vertices labeled 1 and 10 can be joined by an edge in order to form a new super edge-magic graph. We call such a graph an *irregular crown*. Roughly speaking, an irregular crown is a cycle with pendant vertices adjacent to the vertices of the cycle. We call the crown *even* or *odd*, depending on the order of the cycle. The observation made before can be used in order to prove the following result.

Theorem 3.10 *All odd irregular crowns are super edge-magic.*

However, the previous theorem cannot be generalized to even irregular crowns in general. See the next result.

Theorem 3.11 ([9]) *The cycle C_n is super edge-magic if and only if n is odd.*

Proof A cycle of odd order is a particular case of an odd irregular crown. Hence, it is super edge-magic. The fact that C_n is not super edge-magic for n even follows from Corollary 3.4.   □

On the other hand, since all paths are caterpillars, it follows that all paths are super edge-magic.

We already know that all caterpillars are super edge-magic. However, what do we know about trees in general? So far, this is still an open question and Enomoto et al. proposed the following conjecture, that has become very popular.

Conjecture 3.1 ([9]) All trees are super edge-magic.

Observe that a positive answer to this conjecture implies, by Lemma 3.1, a positive answer to the conjecture that states that all trees are harmonious (see Conjecture 4.2).

From this point on, we will concentrate on the edge-magicness and the super edge-magicness of some specific families of graphs, and about what can be said about them.

We already know that the star $K_{1,n}$ is super edge-magic, since it is a particular family of caterpillar. Moreover, we have the following result.

Proposition 3.3 ([11]) *There are exactly $3 \cdot 2^n$ distinct edge-magic labelings of the star $K_{1,n}$, of which only two are super edge-magic, up to isomorphisms.*

Proof Let $V(K_{1,n}) = \{u\} \cup \{v_i : 1 \leq i \leq n\}$ and $E(K_{1,n}) = \{e_i = uv_i : 1 \leq i \leq n\}$. Let f be an edge-magic labeling of $K_{1,n}$ and let σ be its magic sum. Then,

$$\sum_{i=1}^{n}(f(v_i) + f(e_i)) + nf(u) = n\sigma.$$

Thus, n divides the sum $\sum_{i=1}^{n}(f(v_i) + f(e_i))$. Observe that,

$$\sum_{i=1}^{n}(f(v_i) + f(e_i)) + f(u) = \sum_{i=1}^{2n+1} i = 2n^2 + 3n + 1.$$

Hence, n divides the expression $f(u) - 1$ and, since by definition $1 \leq f(u) \leq 2n+1$, we obtain $f(u) \in \{1, n+1, 2n+1\}$. Since $n\sigma = 2n^2 + 3n + 1 + (n-1)f(u)$, it

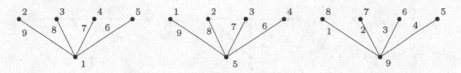

Fig. 3.7 The edge-magic labelings f_i, i=1,2,3 of $K_{1,4}$ introduced in the proof of Proposition 3.3

follows that, $\sigma = 2n+4, 3n+3$ or $4n+2$, which corresponds to $f(u) = 1, n+1$ and $2n + 1$, respectively. At this point, it suffices to exhibit labelings with each of these three possible magic sums, and then describe how to obtain all the other labelings from them. Let f_1, f_2, and f_3 be the edge-magic labelings defined as follows:

$$f_1(u) = 1, \qquad f_1(v_i) = i + 1, f_1(e_i) = 2n + 2 - i,$$
$$f_2(u) = n + 1, \quad f_2(v_i) = i, \qquad f_2(e_i) = 2n + 2 - i,$$
$$f_3(u) = 2n + 1, f_3(v_i) = i, \qquad f_3(e_i) = 2n + 1 - i,$$

where $i \in [1, n]$. Then, the magic sums of f_1, f_2, and f_3 are $2n+4, 3n+3$, and $4n+2$, respectively. Note that, all other edge-magic labelings of $K_{1,n}$ can be obtained by permuting the labels of e_i and v_i for any i with $1 \leq i \leq n$, and that out of these $3 \cdot 2^n$ possible permutations, only f_1 and f_2 are super edge-magic (Fig. 3.7).   □

We want to point out that we immediately obtain the following corollary. We believe that it is interesting since Godbold and Slater [16] have conjectured that for cycles of order different from 5, there are no gaps between the set of magic sums of edge-magic labelings (we will consider this problem with more detail in Sect. 6.6).

Corollary 3.6 ([11]) *For every positive integer $n \geq 2$, there exists a super edge-magic graph G such that $|\sigma_1 - \sigma_2| \geq n-1$, where σ_1 and σ_2 are the only two possible magic sums of G.*

Figure 3.3 shows that $K_2 \vee nK_1$ is super edge-magic. Since $K_2 \vee nK_1 \cong K_{1,1,n}$, we suggest the following problem that we leave as an exercise for the reader.

Exercise 3.9 Characterize the set of complete m-partite super edge-magic graphs.

Next, we consider the family of fans.

Proposition 3.4 ([10]) *The fan $F_n \cong P_n \vee K_1$ is super edge-magic if and only if $1 \leq n \leq 6$.*

Proof Figure 3.8 shows super edge-magic labelings of each F_n, for $n \in [1, 6]$. Let us prove now that these are the only super edge-magic graphs of the family $\{F_n : n \geq 1\}$. Assume to the contrary that there exists $n \in \mathbb{N}$, with $n \geq 7$ such that g is a super-edge magic labeling of F_n. Define $p = n + 1$ and $V(F_n) = \{v_i : g(v_i) = i, i \in [1, p]\}$. Since F_n is super edge-magic and $|E(F_n)| = 2p - 3$, it follows, by Lemma 3.5, that $S = \{g(u) + g(v) : uv \in E(F_n)\} = [3, 2p - 1]$. Now $n \geq 7$, then the vertices $v_1, v_2, v_3, v_4, v_{p-3}, v_{p-2}, v_{p-1}, v_p$ are all mutually distinct.

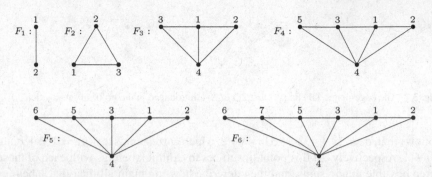

Fig. 3.8 A super edge-magic labeling of F_n, $1 \leq n \leq 6$

Observe that each of $3, 4, 2p - 2, 2p - 1$ can be expressed uniquely as sums of two distinct elements in the set $L = [1, p]$. Namely, $3 = 1 + 2$; $4 = 1 + 3$; $2p - 2 = p + p - 2$; $2p - 1 = p + p - 1$. Therefore, $\{v_1v_2, v_1v_3, v_{p-2}v_p, v_{p-1}v_p\} \subset E(F_n)$. Next, notice that 5 and $2p - 3$ can be each expressed in exactly two ways as sums of distinct elements of L, namely $5 = 1 + 4$ or $2 + 3$, and $2p - 3 = p + (p - 3)$ or $(p - 1) + (p - 2)$.

Thus, $\{v_1v_4, v_pv_{p-3}\}$, $\{v_1v_4, v_{p-1}v_{p-2}\}$, $\{v_2v_3, v_pv_{p-3}\}$, and $\{v_2v_3, v_{p-1}v_{p-2}\}$ are four mutually exclusive possibilities for being subsets of $E(F_n)$. Finally, by adding any of these four pairs of edges, that are necessarily in $E(F_n)$, we obtain forbidden subgraphs of the fan, namely either $2K_{1,3}$, $K_{1,3} \cup K_3$, or $2K_3$.   □

The next results about ladders and generalized prisms have been found independently by Enomoto and Yokomura and by Figueroa et al. [10].

Proposition 3.5 ([10]) *The ladder $L_n \cong P_n \square P_2$ is super edge-magic, if n is odd.*

Proof Let L_n be the ladder with $V(L_n) = \{u_i, v_i : 1 \leq i \leq n\}$ and

$$E(L_n) = \{u_iu_{i+1}, v_iv_{i+1}, u_jv_j : 1 \leq i \leq n - 1, 1 \leq j \leq n\}.$$

Now, consider the following function $f : V(L_n) \to [1, 2n]$ defined by the rule:

$$f(v) = \begin{cases} (i + 1)/2, & \text{if } v = u_i, \ i \text{ is odd and } 1 \leq i \leq n, \\ (n + i + 1)/2, & \text{if } v = u_i, \ i \text{ is even and } 1 \leq i \leq n, \\ (3n + i)/2, & \text{if } v = v_i, \ i \text{ is odd and } 1 \leq i \leq n, \\ (2n + i)/2, & \text{if } v = v_i, \ i \text{ is even and } 1 \leq i \leq n. \end{cases}$$

Thus, an easy check shows that f is a super edge-magic labeling of L_n.   □

The converse of the previous theorem is not true, since, for instance, $L_2 \cong C_4$ is not super edge-magic. However, L_4 and L_6 are super edge-magic. See Fig. 3.9.

This leads to the following question that remains an open problem.

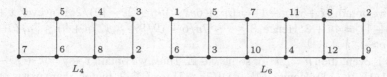

Fig. 3.9 A super edge-magic labeling of L_4 and L_6

Problem 3.1 Characterize the set of even integers n for which the graph L_n is super edge-magic [10].

Proposition 3.6 ([10]) *The generalized prism $Y_{m,n} = C_m \Box P_n$ is super edge-magic if m is odd and $n \geq 2$.*

Proof Define the generalized prism $C_m \times P_n$ as follows. $V(C_m \times P_n) = \{v_{i,j} : 1 \leq i \leq m$ and $1 \leq j \leq n\}$ and $E(C_m \times P_n) = \{v_{i,j}v_{i+1,j}, v_{m,j}v_{1,j} : 1 \leq i \leq m-1, 1 \leq j \leq n\} \cup \{v_{i,j}v_{i,j+1} : 1 \leq i \leq m, 1 \leq j \leq n-1\}$.

Consider the following function $f : V(C_m \times P_n) \to [1, mn]$ where

$$
f(v_{i,j}) = \begin{cases}
m(j-1) + (i+1)/2, & \text{if } i \text{ is odd and } j \text{ is odd,} \\
m(j-1) + (i+m+1)/2, & \text{if } i \text{ is even and } j \text{ is odd,} \\
m(j-1) + i/2, & \text{if } i \text{ is even and } j \text{ is even,} \\
m(j-1) + (i+m)/2, & \text{if } i \text{ is odd and } j \text{ is even.}
\end{cases}
$$

Then, it is easy to check that f is a super edge-magic labeling of $Y_{m,n}$. □

Notice that, by Corollary 3.4, when $n = 2$ and m is even the graph $C_m \times P_2$ is not super edge-magic. For $n > 2$ and m even, the super edge-magicness of $C_m \times P_n$ is, as far as we know, unknown [10].

Next, we study the book $B_n \cong K_{1,n} \Box K_2$.

Proposition 3.7 ([10]) *If the book B_n is super edge-magic, with a super edge-magic labeling f such that $s = \min\{f(x)+f(y) : xy \in E(B_n)\}$ then the following conditions are satisfied:*

(i) *If n is odd, then $n \equiv 5 \pmod 8$ and $s \in \{(n+27)/8, (3n+25)/8, (5n+23)/8, (7n+21)/8, (9n+19)/8\}$ unless $n = 5$, in which case, s can also be 3.*

(ii) *If n is even, then $s = n/2 + 3$ unless $n = 2$, in which case, s can be 3.*

Proof The book B_n has order $p = 2n + 2$ and size $q = 3n + 1$. If x and y represent the labels of the two vertices of degree $n+1$ of B_n, then the sum of all induced edge sums is:

$$
2 \sum_{i=1}^{2n+2} i + (x+y)(n-1) = (3n+1)s + \frac{3n(3n+1)}{2}.
$$

Thus, we obtain $(x + y)(n - 1) = (n^2 + 6ns - 17n + 2s - 12)/2$. However, $x + y \leq p + (p - 1) = 4n + 3$. Hence, $3 \leq s \leq 7n/6 + 19/9 + 8/(27n + 9) \leq 7n/6 + 7/3$, since $n \geq 1$.

If n is even, then $n = 2k$ for some $k \in \mathbb{Z}$. Thus, we obtain $x + y = k + 3s - 8 + (4s - 14)/(2k - 1)$. Hence, $(2k - 1)|(2s - 7)$ for $k \geq 2$, that is, there exists $m \in \mathbb{Z}$ such that $(m(2k - 1) + 7)/2 = s$. Then, we obtain $-1 \leq m \leq 2$, implying that $m = 1$, since $s \in \mathbb{Z}$ and $k \geq 2$. Hence, $s = n/2 + 3$. For $n = 2$, notice that, $s = 3$ or $s = 4$.

For the cases when n is odd, if $n \equiv 3 \pmod 4$, then every vertex of B_n is even and $q \equiv 2 \pmod 4$. Thus B_n is not super edge-magic by Corollary 3.3.

Now, if $n \equiv 1 \pmod 4$, then $n = 4k + 1$ for some $k \in \mathbb{Z}$ and, $2(x + y) = 4k + 6s - 15 + (2s - 7)/k$. Thus, we obtain $k|(2s - 7)$. Now, if $n = 8k - 1$ for some $k \in \mathbb{Z}$, then $2k|(2s - 7)$ which is impossible. Hence, when n is odd there exists $k \in \mathbb{Z}$ such that $(m(2k + 1) - 1)/2 = s$. Then $-1 \leq m \leq 9$, which implies $m \in \{-1, 1, 3, 5, 7, 9\}$. Therefore,

$$s \in \left\{ \frac{-n + 29}{8}, \frac{n + 27}{8}, \frac{3n + 25}{8}, \frac{5n + 23}{8}, \frac{7n + 21}{8}, \frac{9n + 19}{8} \right\}.$$

Finally, notice that $s = (-n + 29)/8$ only when $n = 5$, which completes the proof.

□

As far as we know, the following conjecture is still open.

Conjecture 3.2 ([10]) For every integer $n \geq 5$, the book B_n is super edge-magic if and only if n is even or $n \equiv 5 \pmod 8$.

3.2.2 Nonisomorphic Labelings

Let G be a graph. We say that two vertex labelings f_1 and f_2 of G are isomorphic if there exists an automorphism φ of G such that for every pair of vertices x, y of G, $\varphi(x) = y$ if and only if $f_1(x) = f_2(y)$. In our task to develop tools to study labelings of graphs, we have found out that super edge-magic labelings of graphs with equal order and size are of great importance. In fact, it turns out that it is not only important to know which graphs of equal order and size are super edge-magic, but also to know how many nonisomorphic labelings can be produced. A very important subfamily of such graphs is the family of cycles. A classical result in the area establishes that a cycle is super edge-magic if and only if its order is odd. But how many nonisomorphic super edge-magic labelings of cycles of odd order are there? To find the exact value seems to be a very difficult question in general. Even to find lower bounds for the number of such labelings is not an easy problem. However, we count with a very interesting lower bound on the number of α-valuations of paths, established by Abrham and Kotzig in [1], that allows us to find a lower bound on the number of nonisomorphic super edge-magic labelings of cycles of odd order.

In this section, we will see that the number of SEM labelings of C_n grows exponentially with n. To prove this fact, we will consider the relation existing between SEM labelings of cycles of odd order with some α-valuations of P_n, and the lower bound introduced for them by Abrham and Kotzig.

Let $0 \leq d < n - 1$ and let P_n be a path with $V(P_n) = \{v_i : 1 \leq i \leq n\}$ and $E(P_n) = \{v_i v_{i+1} : 1 \leq i \leq n - 1\}$. Let f be an α-*valuation* of P_n. Then f will be called an α_d-valuation of P_n if $\min\{f(v_1), f(v_n)\} = d$.

Theorem 3.12 ([1]) *Let $N_d(n)$ be the number of nonisomorphic α_d-valuations of P_{n+1}. Then,*

1. *$N_0(n) = 1$, for every $n \geq 1$.*
2. *$N_1(n) \geq 2^{\lfloor n/3 \rfloor - 2}$, for every $n \geq 5$.*
3. *$N_2(n) \geq 2^{\lfloor n/3 \rfloor}$, for every $n \geq 9$.*

By Corollary 3.1, if a tree T admits an α-valuation, then T also admits a super edge-magic labeling. A relationship between α_d-valuations of P_n, for n odd and $d = 0, 1, 2$, and super edge-magic labelings of cycles is established in the next lemma.

Lemma 3.6 ([4]) *Let P_n be a path on n vertices, $n \geq 3$ odd. If P_n admits an α_d-valuation for $d = 0, 1, 2$, then the cycle C_n admits a super edge-magic labeling.*

Proof Abrham and Kotzig [1] proved that if f is an α-valuation of the path P_{2t+1}, and $\min\{f(v_1), f(v_{2t+1})\} \leq t - 1$, then $f(v_1) + f(v_{2t+1}) = t = k$, where k is the characteristic of f. Moreover, the vertices with the values greater than k and the vertices with the values less than or equal to k necessarily alternate.

Let f_d be an α_d-valuation of P_{2t+1}, for $d = 0, 1$ or 2 satisfying $f(v_1) < f(v_{2t+1})$ and $f(v_1) \leq t - 1$. Consider the following labeling of the vertices of P_{2t+1}:

$$g(v_i) = \begin{cases} f(v_i) + 1, & \text{if } i \text{ even,} \\ t + 1 - f(v_i), & \text{if } i \text{ odd.} \end{cases}$$

Notice that, since $f(v_1) \leq t - 1 = k - 1$, the labels assigned by g to the vertices v_i are $[1, k + 1]$, when i is odd; and those assigned to the remaining vertices are $[k + 2, n]$. Thus, the function g is an injection from $V(P_n)$ onto $[1, n]$. Furthermore, we obtain the following equality of sets: $\{|f(v_i) - f(v_{i+1})| : i \in [1, n - 1]\} = [1, n - 1]$, since f is an α-valuation and $\{g(v_i) + g(v_{i+1}) = t + 2 + |f(v_i) - f(v_{i+1})| : i \in [1, n - 1]\} = [t + 3, 3t + 2]$. Thus,

- if $d = 0$, then $f(v_1) = 0, f(v_n) = t, g(v_1) = t + 1$ and $g(v_n) = 1$.
- if $d = 1$, then $f(v_1) = 1, f(v_n) = t - 1, g(v_1) = t$ and $g(v_n) = 2$.
- if $d = 2$, then $f(v_1) = 2, f(v_n) = t - 2, g(v_1) = t - 1$ and $g(v_n) = 3$.

Hence, for each previous case, we obtain the sum $g(v_1) + g(v_n) = t + 2$. Therefore, by Lemma 2.1, the vertex labeling g can be extended to a super edge-magic labeling of the cycle C_n with magic sum $5t + 4$. □

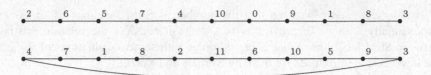

Fig. 3.10 An α_2-valuation of P_{11} and the induced SEM of C_{11} (introduced in Lemma 3.6)

Example 3.1 An α_2-valuation of P_{11} and the super edge-magic labeling of C_{11} obtained from it appear in Fig. 3.10.

The next theorem gives an exponential lower bound for the number of super edge-magic labelings of the cycle C_n, when n is odd.

Theorem 3.13 ([4]) *Let C_n be a cycle on n vertices, $n \geq 11$ odd. The number of nonisomorphic super edge-magic labelings of the cycle C_n is at least $5/4 \cdot 2^{\lfloor (n-1)/3 \rfloor} + 1$.*

Proof By Theorem 3.12 and Lemma 3.6, we have that, for $n \geq 11$ odd, the number of super edge-magic labelings of the cycle C_n is at least $N_0(n-1) + N_1(n-1) + N_2(n-1) \geq 5/4 \cdot 2^{\lfloor (n-1)/3 \rfloor} + 1$. $\square$

The bound obtained in the previous theorem implies the following result, which involves special super edge-magic labelings of matchings of odd size. Notice that, by Corollary 3.4, matchings of even size are not super edge-magic.

Theorem 3.14 *Assume that $n \geq 11$ is odd. Let nK_2 be a matching of size n. Then the number of nonisomorphic special super edge-magic labelings of nK_2 is lower bounded by $5/2 \cdot 2^{\lfloor (n-1)/3 \rfloor} + 2$.*

Proof Let n be odd. We will show that any super edge-magic labeling of any 2-regular graph G on n vertices can be transformed into at least two different super edge-magic labelings of nK_2. Let G be any super edge-magic labeled 2-regular graph of order n, where each vertex takes the name of its label. Since the disjoint union of n loops is not super edge-magic, it follows that G admits at least two distinct orientations of its edges in which each component has been oriented either clockwise or counterclockwise. Let G^+ be one of these orientations and construct its adjacency matrix, namely $A(G^+)$. It is obvious that two distinct oriented 1-regular super edge-magic digraphs will provide distinct adjacency matrices. Next, we transform the $n \times n$ matrix $A(G^+)$ into a $2n \times 2n$ matrix $B(G^+)$ as follows. We consider a square null matrix of order $2n$ and we replace the $n \times n$ submatrix generated for the last n columns and the first n rows by $A(G^+)$. Then, $B(G^+)$ is the adjacency matrix of a directed special super edge-magic matching. Notice that if two matrices $B(G^+)$ and $B(G'^+)$ are distinct then the special super edge-magic labelings of the underlying matchings are distinct, and therefore, by Theorem 3.13 the result follows. $\square$

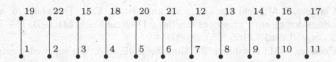

Fig. 3.11 A super edge-magic labeling of $11K_2$

It is worthwhile mentioning that the idea used in this proof has been used also in [7] and [20].

Example 3.2 Figure 3.11 shows one of the two possible SEM labeling of $11K_2$ obtained from the labeling of C_{11} in Fig. 3.10 (orienting the cycle in such a way that $(4, 7)$ is an arc of the digraph and the resulting digraph is strongly connected), using the construction explained in the proof of Theorem 3.14.

In Chap. 6, new techniques will be introduced that will allow us to improve the bounds obtained in this chapter for many different values of n.

Acknowledgements We gratefully acknowledge permission to use [2, 3, 11, 17] by the publisher of Ars Combinatoria and Utilitas Mathematica. We also gratefully acknowledge permission to use [9, 13] by the authors and publisher of SUT J. Math. and J. Comb. Math. and Comb. Comput. The proofs from [10] are introduced with permission from [10], Elsevier, ©2001. The proofs from [4] are introduced with permission from Springer, [4], ©Institute of Mathematics, Academy of Mathematics and Systems Science, Chinese Academy of Sciences, Chinese Mathematical Society and Springer Berlin Heidelberg 2009.

References

1. Abrham, J., Kotzig, A.: Exponential lower bounds for the number of graceful numbering of snakes. Congr. Numer. **72**, 163–174 (1990)
2. Bača, M., Barrientos, C.: Graceful and edge-antimagic labelings. Ars Comb. **96**, 505–513 (2010)
3. Bača, M., Lin, Y., Miller, M., Simanjuntak, R.: New constructions of magic and antimagic graph labelings. Utilitas Math. **60**, 229–239 (2001)
4. Bača, M., Lin, Y., Muntaner-Batle, F.A., Rius-Font, M.: Strong labelings of linear forests. Acta Math. Sin. Engl. Ser. **25**(12), 1951–1964 (2009). http://dx.doi.org/10.1007/s10114-009-8284-3
5. Bača, M., Kováč, P., Semaničová-Feňovčíková, A., Shafiq, M.K.: On super $(a, 1)$-edge-antimagic total labeling of regular graphs. Discrete Math. **310**, 1408–1412 (2010)
6. Bača, M., Lascáková, M., Semaničová, A.: On connection between α-labelings and edge antimagic labelings of disconnected graphs. Ars Comb. **106**, 321–336 (2012)
7. Bloom, G., Lampis, M., Muntaner-Batle, F.A., Rius-Font, M.: Queen labelings. AKCE Int. J. Graphs Comb. **8**(1), 13–22 (2011)
8. Cairnie, N., Edwards, K.: The computational complexity of cordial and equitable labeling. Discrete Math. **216**, 29–34 (2000)
9. Enomoto, H., Lladó, A., Nakamigawa, T., Ringel, G.: Super edge-magic graphs. SUT J. Math. **34**, 105–109 (1998)

10. Figueroa-Centeno, R.M., Ichishima, R., Muntaner-Batle, F.A.: The place of super edge-magic labelings among other classes of labelings. Discrete Math. **231**(1–3), 153–168 (2001). http://dx.doi.org/10.1016/S0012-365X(00)00314-9
11. Figueroa-Centeno, R.M., Ichishima, R., Muntaner-Batle, F.A.: On super edge-magic graphs. Ars Comb. **64**, 81–95 (2002)
12. Figueroa-Centeno, R.M., Ichishima, R., Muntaner-Batle, F.A.: On edge magic labelings of certain disjoint unions of graphs. Australas. J. Comb. **32**, 225–242 (2005)
13. Figueroa-Centeno, R.M., Ichishima, R., Muntaner-Batle, F.A., Rius-Font, M.: Labeling generating matrices. J. Comb. Math. Comb. Comput. **67**, 189–216 (2008)
14. Gallian, J.A.: A guide to the graph labeling zoo. Discrete Appl. Math. **49**(1–3), 213–229 (1994)
15. Gallian, J.A.: A dynamic survey of graph labeling. Electron. J. Comb. **19**(DS6) (2016)
16. Godbold, R.D., Slater, P.J.: All cycles are edge-magic. Bull. Inst. Comb. Appl. **22**, 93–97 (1998)
17. Ichishima, R., Muntaner-Batle, F.A., Rius-Font, M.: Bounds on the size of super edge-magic graphs depending on the girth. Ars Comb. **119**, 129–133 (2015)
18. Kotzig, A., Rosa, A.: Magic valuations of finite graphs. Can. Math. Bull. **13**, 451–461 (1970)
19. Kotzig, A., Rosa, A.: Magic valuations of complete graphs. Publ. CRM **175** (1972)
20. Muntaner-Batle, F.A.: Special super edge-magic labelings of bipartite graphs. J. Comb. Math. Comb. Comput. **39**, 107–120 (2001)
21. Rosa, A.: On certain valuations of the vertices of a graph. In: Theory of Graphs (Internat. Symposium, Rome, July 1966), pp. 349–355. Gordon and Breach/Dunod, New York/Paris (1967)

Chapter 4
Harmonious Labelings

4.1 Introduction

Harmonious labelings are very important in the literature, since many authors have
devoted their efforts to better understanding them. Thus, we believe that it is worth
the while to dedicate special attention to them. However many of the results that
involve harmonious labelings can be obtained using similar ideas to the ones used
for super edge-magic labelings as well as using the relations that appear in Sects. 3.1
and 6.3. Therefore, we will dedicate only a short chapter to this important labeling.
We start answering the question of which complete graphs are harmonious. Later
we will study a few families of harmonious graphs as well as general properties of
these graphs. We will conclude this chapter providing an asymptotic answer to the
question of how many graphs are harmonious.

For integers $m \leq n$, we denote the set $\{m, m + 1, ..., n\}$ by $[m, n]$.

4.2 Harmonious Labelings of Complete Graphs

One possible non-standard way to attack the problem of the harmonicity of complete
graphs is by means of what is called by Ruzsa [12] weak Sidon sets or equivalently,
by Kotzig [7] well-spread sets. Weak Sidon sets are motivated by Sidon sets,
introduced by the Hungarian mathematician Simon Sidon in [13].

Definition 4.1 A set $S = \{a_1, a_2, \ldots, a_n\} \subset \mathbb{Z}$ is a *Sidon set* of cardinality n if all
pairwise sums $a_i + a_j$, $i \leq j$, are distinct.

Example 4.1 The set $A = \{0, 1, 3, 7\}$ is a Sidon set of cardinality 4.

Definition 4.2 ([7, 12]) Let $A = \{a_1, a_2, \ldots, a_n\}$ be a subset of nonnegative integers
such that for any four distinct elements of A, namely a_i, a_j, a_k, a_l, we have $a_i + a_j \neq
a_k + a_l$. Then A is called a *weak Sidon set* or a *well-spread set* of integers. Thus, A

© The Author(s) 2017
S.C. López, F.A. Muntaner-Batle, *Graceful, Harmonious and Magic Type Labelings*,
SpringerBriefs in Mathematics, DOI 10.1007/978-3-319-52657-7_4

is a weak Sidon set or a well-spread set if and only if $|2^\wedge A| = n(n-1)/2$, where $2^\wedge A = \{a + b : a, b \in A \text{ and } a \neq b\}$.

Example 4.2 The set $A = \{1, 2, 3, 5\}$ is a well-spread set since $2^\wedge A = [3, 8]$, thus $|2^\wedge A| = 6$. On the other hand, the set $B = \{1, 2, 3, 4\}$ is not a well-spread set since $|2^\wedge B| = 5$.

Notice that the labels of the vertices of any harmonius labeling of the complete graph K_n (if we consider the smallest nonnegative integer representing each class) must form a weak Sidon subset of cardinality n of the set $[0, \binom{n}{2} - 1]$. Otherwise, there exist four distinct vertices of K_n, namely x, y, x', y' such that $f(x) + f(y) = f(x') + f(y')$. But, this implies that $f(x) + f(y) \equiv f(x') + f(y') \pmod{\binom{n}{2}}$. Hence, f is not a harmonious labeling of K_n. Therefore, a natural question to ask is the following one.

For which values of n, is the set $[0, \binom{n}{2} - 1]$ big enough in order to contain a weak Sidon set A of cardinality n?

In order to partially answer this question, let us introduce the concepts of *span of a weak Sidon set* of cardinality n, and of span of n.

Definition 4.3 ([7]) Let A be a weak Sidon set of cardinality n. The span of A, denoted by $\sigma(A)$, is defined to be

$$\sigma(A) = \max A - \min A + 1.$$

The *span of* n, denoted by $\sigma^*(n)$, is

$$\sigma^*(n) = \min\{\sigma(A) : A \text{ is a weak Sidon set with } |A| = n\}.$$

The next list taken from [7] shows the values of $\sigma^*(n)$ for all values of $n \in [3, 12]$.

$$\sigma^*(3) = 3, \quad \sigma^*(4) = 5, \quad \sigma^*(5) = 8, \quad \sigma^*(6) = 13, \quad \sigma^*(7) = 19,$$
$$\sigma^*(8) = 25, \sigma^*(9) = 35, \sigma^*(10) = 46, \sigma^*(11) = 58, \sigma^*(12) = 72.$$

It is clear that for $n \in [3, 9]$, we have $\sigma^*(n) < \binom{n}{2}$. Therefore, the set $[0, \binom{n}{2} - 1]$ may contain weak Sidon sets of cardinality n. However, for $n = 10, 11, 12$, we have $\sigma^*(n) > \binom{n}{2}$. Therefore, the set $[0, \binom{n}{2} - 1]$ does not contain weak Sidon sets of cardinality n. Hence, as an immediate corollary, we obtain that the graphs K_{10}, K_{11}, and K_{12} are not harmonious.

We are not aware of any other known values of $\sigma^*(n)$ for $n \geq 13, n \in \mathbb{Z}$. As in the case of Sidon sets, they are probably very hard to be calculated. However, we highly suspect that for these values, $\sigma^*(n) > n(n-1)/2$. This would be enough to eliminate all graphs K_n ($n \geq 10, n \in \mathbb{Z}$) from being harmonious. Also, the bounds that we have so far for weak Sidon sets are just not good enough to eliminate the cases from $n = 13, 14, \ldots, 197$. However, for $n \geq 198$, we can use the following result by Ruzsa [12], to show that K_n is not harmonious.

Theorem 4.1 ([12]) *Let A be any weak Sidon set with* $\min A = 0$ *and* $\max A = N$. *Then*

$$|A| < \sqrt{N} + 4\sqrt[4]{N} + 11.$$

We want to see for which values of n, the following inequality is true.

$$n > \sqrt{\frac{n(n-1)}{2} - 1} + 4\sqrt[4]{\frac{n(n-1)}{2} - 1} + 11. \tag{4.1}$$

Easy calculations show that for $n \geq 198$, (4.1) is satisfied, and hence we have the following result.

Theorem 4.2 *The graph* K_n *is not harmonious when* $n = 10, 11, 12$ *or* $n \geq 198$.

The remaining values of n have to be checked case by case, when using this approach. For some specific values of n, we will be able to prove that K_n is not harmonious using results that will be stated further in this section. One of these results is Theorem 4.3. In order to prove this result, we follow an idea found in [14].

Notice that, if q is even, then the parity of each element of $\mathbb{Z}_q$ is well defined and the sum of all elements in $\mathbb{Z}_q$ is

$$1 + 2 + \ldots + \frac{q}{2} + \ldots + q - 2 + q - 1 \equiv \frac{q}{2} \pmod{q}.$$

Theorem 4.3 ([14]) *Let* $n \geq 4$ *be a positive integer such that* $n \equiv 0$ *or* $1 \pmod 4$. *If* K_n *is harmonious, then* n *is a perfect square.*

Proof Let n be a positive integer. Suppose there exists a harmonious labeling of K_n, namely $f : V(K_n) \to \mathbb{Z}_m$, where $m = \binom{n}{2}$. For every $xy \in E(K_n)$, let $g(xy) = f(x) + f(y) \pmod m$. By definition of a harmonious labeling of K_n, it is clear that

$$S = \{g(xy) : xy \in E(K_n)\} = \mathbb{Z}_m. \tag{4.2}$$

Since m is even, we can consider the following partitions of vertices and edges. Let $V_e = \{x \in V(K_n) : f(x) \text{ is even}\}$ and $V_o = V(K_n) \setminus V_e$. Consider the partition of edges $E(K_n) = E_1 \cup E_2$, where $E_1 = \{xy : x, y \in V_e\} \cup \{xy : x, y \in V_o\}$ and $E_2 = E(K_n) \setminus E_1$. In particular, we have $g(xy)$ is even if and only if $xy \in E_1$. Let $p = |V_e|$. Then, $|E_1| = \binom{p}{2} + \binom{n-p}{2}$ and $|E_2| = p(n-p)$.

By (4.2), it follows that $|E_1| = |E_2|$. Thus, $\binom{p}{2} + \binom{n-p}{2} = p(n-p)$, which is equivalent to $4p^2 - 4np + n^2 - n = 0$. Hence, we obtain $p = (n \pm \sqrt{n})/2$. Therefore, n must be a perfect square. $\square$

By analyzing the values that remain from Theorems 4.2 and 4.3, there are still many cases to be checked one by one. However, if we do it, then we will obtain that the only harmonious graphs are K_1, K_2, K_3, and K_4. Figure 4.1 shows harmonious labelings for these complete graphs.

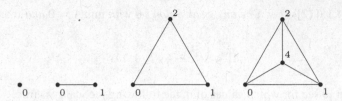

Fig. 4.1 A harmonious labeling of K_n, for $n = 1, 2, 3$ and 4

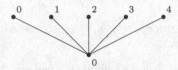

Fig. 4.2 An example of a harmonious star

Therefore we have the following theorem found first in [5].

Theorem 4.4 ([5]) *The complete graph K_n is harmonious if and only if $n \leq 4$.*

We want to mention that the original proof found in [5] follows a slightly different approach. We conclude this section with the following open problem.

Problem 4.1 Determine the exact value of $\sigma^*(n)$, for all values of n or at least for new values of n.

4.3 Other Harmonious Graphs

The family of complete graph has been considered in the previous section. Another interesting family to consider is the family of complete bipartite graphs. The next result characterizes the harmonious complete bipartite graphs.

Theorem 4.5 ([5]) *The complete bipartite graph $K_{m,n}$ is harmonious if and only if m or n is equal to 1.*

Proof The graph $K_{1,n}$ is harmonious. It suffices to label the center vertex of $K_{1,n}$ with 0 and the rest of the vertices in any way with all numbers in $[0, n-1]$, see, for instance, Fig. 4.2. Next we have to show that if $\min\{m, n\} \geq 2$, then $K_{m,n}$ is not harmonious. Assume to the contrary that $K_{m,n}$ is harmonious. Then, there exist $A, B \subset \mathbb{Z}_{mn}$ such that $|A| = m$, $|B| = n$ and $A \cap B = \emptyset$. By definition of a harmonious labeling, each element of $\mathbb{Z}_{mn}$ can be expressed in a unique way as the sum $a + b \pmod{mn}$, where $a \in A$ and $b \in B$, and thus all these sums are distinct. Thus, all the differences $a - b \pmod{mn}$ are distinct. But there are mn differences and $|\mathbb{Z}_{mn}| = mn$. Hence, there is a difference $a - b = 0$ in $\mathbb{Z}_{mn}$. That is, although we are assuming that $A \cap B \neq \emptyset$, we get that $a = b$, a contradiction. Therefore, the graph $K_{m,n}$ is not harmonious.   □

At this point it is interesting to discuss a problem that appeared in 2001, and that has remained opened until 2012 in spite of many efforts. The first complete proof that appeared in the literature is found in [10]. However, it is somewhat unpleasant to read. The proof that we propose in this book is due to Figueroa-Centeno and Ichishima [6] and their solution is highly linked to Theorem 4.5 and the relations established in Chap. 3. In fact, this result is a good example of the power of these relations.

The *edge-magic deficiency* of a graph G, $\mu(G)$, is defined to be the minimum positive integer n such that $G \cup nK_1$ is an edge-magic graph. This definition was first introduced by Kotzig and Rosa [7, 8], who showed that $\mu(G)$ was well defined. This motivates Figueroa-Centeno et al. to introduce the concept of super edge-magic deficiency found in [2].

Definition 4.4 ([2]) Let G be a (p, q)-graph. The set $S(G)$ is

$$S(G) = \{n \in \mathbb{N} \cup \{0\} : G \cup nK_1 \text{ is super edge-magic}\}.$$

Then, the *super edge-magic deficiency* of G, denoted by $\mu_s(G)$ is defined as

$$\mu_s(G) = \begin{cases} \min S(G), & \text{if } S(G) \neq \emptyset, \\ +\infty, & \text{otherwise.} \end{cases} \tag{4.3}$$

Figueroa-Centeno et al. conjectured in [2] the following result that was first proved in [10] and later in [6].

Theorem 4.6 *Let m and n be positive integers. Then $\mu_s(K_{m,n}) = (m - 1)(n - 1)$.*

Proof ([6]) First we show that $\mu_s(K_{m,n}) \leq (m-1)(n-1)$. Let $V(K_{m,n} \cup (m-1)(n-1)K_1) = \{x_i\}_{i=1}^{m} \cup \{y_j\}_{j=1}^{n} \cup \{z_k\}_{k=1}^{(m-1)(n-1)}$ and $E(K_{m,n} \cup (m-1)(n-1)K_1) = \{x_iy_j : 1 \leq i \leq m, \ 1 \leq j \leq n\}$. Then, the labeling $f : V(K_{m,n} \cup (m-1)(n-1)K_1) \to [1, (m-1)(n-1) + m + n]$ defined by the rule

$$f(v) = \begin{cases} i, & \text{if } v = x_i, \\ m + 1, & \text{if } v = y_1, \\ m + 1 + m(i - 1), & \text{if } v = y_j, \ 2 \leq j \leq n, \end{cases} \tag{4.4}$$

and that assigns the remaining labels to label the vertices of the set $\{z_k\}_{k=1}^{(m-1)(n-1)}$ is a super edge-magic labeling of $K_{m,n} \cup (m - 1)(n - 1)K_1$. Thus, it follows that $\mu_s(K_{m,n}) \leq (m - 1)(n - 1)$. Next, we show that $\mu_s(K_{m,n}) \geq (m - 1)(n - 1)$. By Lemma 3.1, if G is a super edge-magic (p, q)-graph, then G is harmonious whenever $q \geq p$ or G is a tree. Hence, taking the contrapositive we get that when G is a graph with $q \geq p$ or it is a tree, and G is not harmonious, then G is not super edge-magic.

From Theorem 4.5, it is clear that none of the graphs $K_{m,n}$, $K_{m,n} \cup K_1$, $K_{m,n} \cup 2K_1$, $\ldots$, $K_{m,n} \cup ((m - 1)(n - 1) - 1)K_1$ are harmonious, and hence, they are not super edge-magic. Thus, $\mu_s(K_{m,n}) \geq (m-1)(n-1)$. Therefore, we have that $\mu_s(K_{m,n}) = (m-1)(n-1)$.   □

Before going back to harmonious labelings, we introduce the following conjecture and exercise.

Conjecture 4.1 ([2]) Let F be a forest with two components. Then $\mu_s(F) \leq 1$.

The purpose of the following exercise is to show that $\mu_s(F)$ can be 1, when F is a forest with two components.

Exercise 4.1 Find, if possible, an example of a forest with exactly two-components with super edge-magic deficiency 1.

Now, we go back to harmonious labelings with the following exercise.

Exercise 4.2 Is the Petersen graph harmonious? [5]

Next we discuss the harmonicity of cycles.

Theorem 4.7 ([5]) *The cycle C_n is harmonious if and only if n is odd.*

Proof Let $V(C_n) = \{v_i\}_{i=1}^n$ and $E(C_n) = \{v_iv_{i+1}\}_{i=1}^{n-1} \cup \{v_nv_1\}$. If n is odd, then the labeling $f : V(C_n) \to \mathbb{Z}_n$ defined by the rule $f(v_i) = i - 1$, for all $i \in [1, n]$ is a harmonious labeling of C_n (Fig. 4.3). Assume now that n is even, $n = 2m$ and that there exists a harmonious labeling f of C_n such that $f(v_i) = a_i$, for $i \in [1, n]$. Now, it is clear that $\mathbb{Z}_n = \{a_1 + a_2, a_2 + a_3, \ldots, a_n + a_1\}$ where all these sums are taken modulo n. Then, when considering the sum of the induced edge labels, each vertex label appears twice. Moreover, since the induced edge labels are distinct we obtain:

$$2 \sum_{i=0}^{n-1} i \equiv k \quad (\text{mod } 2m)$$

$$\sum_{i=0}^{n-1} i \equiv k \quad (\text{mod } 2m).$$

That is, $\sum_{i=0}^{n-1} i \equiv 0 \pmod{2m}$. Hence, $m \equiv 0 \pmod{2m}$, a contradiction. Therefore, the cycle C_n is not harmonious when n is even.   □

Fig. 4.3 An example of a harmonious cycle

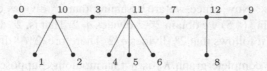

Fig. 4.4 An example of a harmonious caterpillar

Next we introduce the following exercise, which has a strong relation with Theorem 4.7 and constitutes a curious application of harmonious labelings to geometry.

Exercise 4.3 Place n dots in the plane forming the vertices of a regular polygon of n vertices. Show that we can embed the cycle C_n in this scheme with all edges being straight segments and no two edges being parallel if and only if n is odd [5].

Exercise 4.4 Prove that all caterpillars are harmonious. (Hint: See Fig. 4.4.) [5]

In fact, Exercise 4.4 is a special case of what it is suspected to be a much more general result, conjectured by Graham and Sloane in [5], and that nowadays is still unsolved in spite of many attempts.

Conjecture 4.2 All trees are harmonious.

So far, several families of trees have been proved to be harmonious. However a final solution seems to be far away. Even to know whether lobsters are harmonious is still an open problem. Up to this point, it has been checked by computer that the conjecture is true for every tree of order at most 26 [1] (see [3]).

4.4 General Properties

In this section we will study general properties of harmonious graphs that will help the reader to further understand the nature of these graphs. The next two results will be devoted to provide conditions that will eliminate graphs from being harmonious.

Theorem 4.8 ([5]) *Assume that G is any harmonious (p, q)-graph with q even. If 2^α divides the degree of each vertex of $V(G)$, then $2^{\alpha+1}$ divides q.*

Proof By the hand-shaking lemma, $\sum_{x \in V(G)} d(x) = 2q$. Thus, if we assume that 2^α divides the degree of each vertex of $V(G)$ then $2^{\alpha-1}$ divides q. Let $f : V(G) \to \mathbb{Z}_q$ be any harmonious labeling of G. By definition, $\{f(x) + f(y) \ (\mathrm{mod}\ q) : xy \in E(G)\} = \mathbb{Z}_q$. Hence, the sum of the edge labels is $\sum_{x \in V(G)} d(x)f(x) \equiv 0 + 1 + \ldots + q - 1 \ (\mathrm{mod}\ q)$. But, since q is even, $0 + 1 + \ldots + q - 1 \equiv q/2 \ (\mathrm{mod}\ q)$. This implies that

$$\sum_{x \in V(G)} d(x)f(x) - \lambda q = \frac{q}{2}, \tag{4.5}$$

for some integer λ. Now, since we are assuming that 2^α divides each degree and $2^{\alpha-1}$ divides q, using (4.5) we obtain $2^{\alpha-1}$ divides $q/2$. That is, 2^α divides q. Hence, again using (4.5), it follows that 2^α divides $q/2$. Therefore, $2^{\alpha+1}$ divides q. $\square$

Example 4.3 The complete graph K_9 is not harmonious. Suppose to the contrary that K_9 is harmonious. Obviously, K_9 is a regular graph in which each vertex has degree $8 = 2^3$. Therefore, by Theorem 4.8, 2^4 divides $|E(K_9)| = 36$, a contradiction. Therefore K_9 is not harmonious.

As a particular case of Theorem 4.8 we obtain the following corollary.

Corollary 4.1 *Let G be any (p,q)-graph such that $q \equiv 2$ (mod 4) and each vertex has even degree. Then G is not harmonious.*

Next, we state and prove the following theorem, which is in fact a way of generalizing Theorem 4.3.

Theorem 4.9 *Let G be a harmonious graph of even size q. Then there exists a partition of $V(G)$ into two sets A and B such that the number of edges joining the vertices of A and B is $q/2$.*

Proof Let G be a harmonious graph of even size q and let $f : V(G) \to \mathbb{Z}_q$ be a harmonious labeling of G. Then, $\{f(x) + f(y) \pmod{q} : xy \in E(G)\} = \mathbb{Z}_q$. Since q is even, it follows that there are exactly $q/2$ odd elements in $\mathbb{Z}_q$. These elements must appear as edge induced labels by f on the elements of $E(G)$. But the only way in which this can happen is by adding two labels of different parity. Therefore, if we let $A = \{x \in V(G) : f(x) \text{ is odd}\}$ and $B = \{x \in V(G) : f(x) \text{ is even}\}$ then, we are done. $\square$

Example 4.4 The complete graph K_{12} is not harmonious. Suppose to the contrary that K_{12} is harmonious. By Theorem 4.9, there exists a partition of $V(K_{12})$ into two sets such that the number of edges joining vertices in different sets is $|E(K_{12})|/2 = 33$. On the other hand, for any partition $A \cup B$ of the vertex set, the number of edges joining the vertices of A with the vertices of B is $|A||B|$. By exploring the different possible partitions we obtain the next table:

$|A|$	1	2	3	4	5	6
$|B|$	11	10	9	8	7	6
$|A||B|$	11	20	27	32	35	36

Thus, since none of these products equals 33, a contradiction follows. $\square$

Exercise 4.5 Prove the following result: "if G is any (p,q)-graph such that $q \equiv 2$ (mod 4) and each vertex has even degree then there is not any partition of $V(G)$ into two sets A and B such that the number of edges joining A and B is equal to $q/2$."

(Hint: See Exercise 1.7.)

4.5 How Many Graphs Are Harmonious?

Although many graphs have been proven to be harmonious, the property of a graph being harmonious is not common among the set of graphs. In fact, using a similar idea of a proof of Erdős refereeing to the probability of a graph being graceful, Graham and Sloan [5] obtained the following result.

Theorem 4.10 ([5]) *Almost all graphs are not harmonious.*

Proof For our model of a random graph with n vertices we assume that each of the $\binom{n}{2}$ possible edges independently exists or does not with probably $1/2$. Fix $\epsilon \in (0, 1/2)$, and let m be a fixed integer in the range $[(1/2-\epsilon)\binom{n}{2}, (1/2+\epsilon)\binom{n}{2}]$. We shall show that almost no graphs with n vertices and m edges are harmonious, as n goes to infinity. Since almost all graphs with n vertices have a number of edges in this range, the theorem follows.

There are $\binom{n(n-1)/2}{m}$ labeled graphs with n vertices and m edges, and so at least

$$\frac{1}{n!}\binom{n(n-1)/2}{m}$$

unlabeled graphs with n vertices and m edges.

Let f be a labeling of the n vertices with distinct numbers from $\{0, 1, \ldots, m-1\}$. There are $m(m-1)\ldots(m-n+1) \le m^n$ of such labelings. Let us consider how many graphs there are for which f is a harmonious labeling. Let p_i be the number of pairs of vertices $\{v, v'\}$ with $f(v) + f(v') \equiv i \pmod{m}$. Then

$$\sum_{i=0}^{m-1} p_i = \binom{n}{2}.$$

A graph with this labeling is harmonious if it consists of one edge taken from each of the classes counted by p_i. Thus there are

$$\prod_{i=0}^{m-1} p_i$$

labeled graphs for which f is a harmonious labeling. This product is maximized by taking the p_i's as equal as possible. In particular

$$\prod_{i=0}^{m-1} p_i \le \left(\frac{n(n-1)}{2m}\right)^m.$$

Therefore there are at most

$$m^n \left(\frac{n(n-1)}{2m} \right)^m$$

harmonious labeled graphs. This is also an upper bound on the number of harmonious unlabeled graphs. To complete the proof we show that the ratio

$$\rho = \frac{m^n \left(\frac{n(n-1)}{2m} \right)^m}{\frac{1}{n!} \binom{n(n-1)/2}{m}}$$

goes to 0 when $n \to \infty$ and m is in the required range. Write $m = (1/2 - \mu) \binom{n}{2}$, with $|\mu| \leq 1/2$. Then

$$\rho < \frac{m^n n! \sqrt{8 \binom{n}{2} (\frac{1}{2} - \mu)(\frac{1}{2} + \mu)}}{(\frac{1}{2} - \mu)^m 2^{\binom{n}{2}} H_2(\frac{1}{2} - \mu)},$$

where $H_2(x) \doteq -x \log_2 x - (1 - x) \log_2 (1 - x)$ [11, p. 309]. The denominator is equal to

$$2^{-\binom{n}{2} (\frac{1}{2} + \mu) \log_2(\frac{1}{2} + \mu)}$$

and so $\rho \to 0$ as $n \to \infty$.   □

With Theorem 4.10 in hand, we are now ready to prove Theorem 3.8.

Proof of Theorem 3.8. A well-known result by Gilbert states that almost all graphs are connected [4]. This implies that almost all (p, q)-graphs satisfy either that $q \geq p$ or that the graph is a tree. If we combine this with Theorem 4.10 and Lemma 3.1, we are done.   □

The problem of determining which graphs are harmonious has proven to be a difficult one. It was shown by Auparajita, Dulawat, and Rathore in [9] in 2001 that this is an NP-complete problem. In spite of this, some general results have appeared in the literature. Among them we want to mention the following result found by Youssef, in [15].

Theorem 4.11 ([15]) *The graph nG formed by n disjoint copies of G is harmonious when the following three conditions hold.*

  (i) $|V(G)| \leq |E(G)|$,
 (ii) n is odd,
(iii) G is harmonious.

Acknowledgements The proofs from [5] are introduced with permission from [5], ©1980 Society for Industrial and Applied Mathematics. We gratefully acknowledge permission to use [6] by the publisher of Ars Combinatoria.

References

1. Aldred, R.E.L., Mckay, B.D.: Graceful and harmonius labelings of trees. Bull. Inst. Comb. Appl. **23**, 69–72 (1998)
2. Figueroa-Centeno, R., Ichishima, R., Muntaner-Batle, F.A.: Some new results on the super edge-magic deficiency of graphs. J. Comb. Math. Comb. Comput. **55**, 17–31 (2005)
3. Gallian, J.A.: A dynamic survey of graph labeling. Electron. J. Comb. **19**(DS6) (2016)
4. Gilbert, E.N.: Random graphs. Ann. Math. Stat. **30**, 1141–1144 (1959). http://dx.doi.org/10.1007/s10114-009-8284-3
5. Graham, R.L., Sloane, N.J.A.: On additive bases and harmonious graphs. SIAM J. Algebraic Discret. Methods **1**, 382–404 (1980)
6. Ichishima, R., Oshima, A.: On the super edge-magic deficiency and α-valuations of graphs. Ars Comb. **129**, 157–163 (2016)
7. Kotzig, A.: On well spread sets of integers. Publ. CRM **161** (1972)
8. Kotzig, A., Rosa, A.: Magic valuations of complete graphs. Publi. CRM **175** (1972)
9. Krishnaa, A.K., Dulawat, M.S., Rathore, G.S.: Computational complexity in decision problems. In: Conf. of Raj. Parishad, Udaipur, India, 14–15 December 2001
10. Liu, C.H., Wang, T.M., Char, M.I.: On arithmetic deficiency for bicliques. In: Systems and Informatics (ICSAI), 2012 International Conference on Systems and Informatics, pp. 235–238 (2012). http://dx.doi.org/10.1016/S0012-365X(00)00314-9
11. MacWilliams, F.J., Sloane, N.J.A.: The Theory of Error-Correcting Codes. North-Holland, Amsterdam (1977)
12. Ruzsa, I.Z.: Solving linear equations in a set of integers. Acta Arith. **LXV.3**, 259–282 (1993)
13. Sidon, S.: Ein statz über trigonometrishe polynome und seine anwendung in der thorie der Fourier.Reihen. Math. Ann. **106**, 536–539 (1932)
14. Valdés, L.: Edge-magic K_p. Thirty-second South-Eastern international Conference on Combinatorics, Graph Theory and Computing, Baton Rouge, LA, vol. 153, pp. 107–111 (2001)
15. Youssef, M.Z.: Two general results on harmonious labelings. Ars Comb. **68**, 225–230 (2003)

Chapter 5
Graceful Labelings: The Shifting Technique

5.1 Introduction

Graceful labelings of graphs appeared in 1967 due to the relationship found with the problem of decompositions of graphs, in particular with the problem of decomposing complete graphs into copies of a given tree. Strong relations between graceful labelings and Golomb rulers (which are a different way to understand Sidon sets) were also found.

In spite of the fact that graceful labelings appeared almost 50 years ago, not many general techniques are known in order to generate graceful labelings of graphs. In particular the famous Ringel–Kotzig conjecture (or graceful tree conjecture) which states that all trees are graceful remains open in the present. Unfortunately, even when the graphs considered are restricted to be trees, not many general techniques arc known to generate graceful labelings. Basically all techniques known consist in shifting either an appropriate edge or an appropriate pair of edges. In this way, several trees have been known to be graceful. However, the majority of the papers concentrate their efforts in creating families of graceful graphs and almost no techniques are known in order to identify graphs which are not graceful. It is common to find techniques in this last direction relying on parity conditions, or the density of the graphs, that is to say, the size of the graph compared to its order (see [20]).

In this chapter we study the relations existing between graceful labelings with graph decompositions and Golomb rulers. In addition, we show some techniques to generate graceful graphs and negative results. Some examples are also considered.

Next, we start by introducing Ringel's conjecture on decompositions of complete graphs.

Conjecture 5.1 ([35]) The graph K_{2n+1} can be decomposed into $2n + 1$ subgraphs that all are isomorphic to a given tree with n edges.

© The Author(s) 2017
S.C. López, F.A. Muntaner-Batle, *Graceful, Harmonious and Magic Type Labelings*,
SpringerBriefs in Mathematics, DOI 10.1007/978-3-319-52657-7_5

As a way of attacking Ringel's conjecture, Rosa proposed in 1967 [37] the concept of β-valuation. The relationship existing among β-valuations and Conjecture 5.1 is provided next.

Theorem 5.1 *Let T be a tree of order n. If T is graceful, then K_{2n+1} decomposes into $2n + 1$ isomorphic copies of T.*

Proof We draw the vertices of K_{2n+1} in the plane in such a way that they form the vertices of a regular $(2n + 1)$-gon. We label these vertices with the elements of the additive group $\mathbb{Z}_{2n+1}$ in an increasing way following the clockwise sense. We define the displacement between two vertices to be the minimum number of unit movements needed either in the clockwise or in the counterclockwise sense to get from one vertex to another. It is clear that the maximum displacement among the $2n + 1$ vertices is at most n.

It is also clear that the edges of K_{2n+1} can be partitioned into n displacement classes, each of which has cardinality $2n + 1$. Assume that T is a graceful tree of size n, and embed T in this configuration of vertices that we have just defined above, according to any graceful labeling of T. Now, we define a tree T_i for all $i \in \mathbb{Z}_{2n+1}$, in such a way that each $T_i \cong T$, has $V(T_i) = \{i \pmod{2n + 1}, i + 1 \pmod{2n + 1}, \ldots, i + n \pmod{2n + 1}\}$ and T_i is embedded in our configuration such that $(i + \alpha)(i + \beta) \in E(T_i)$ if and only if $\alpha\beta \in E(T)$.

It is clear that T_0 is just identical to T, and of course T has exactly one edge in each of the displacements classes. To get T_1, we shift each edge of T_0 to the next edge in each of the displacements classes. In general T_i moves i edges, each edge of T_0 in each displacement class. Hence, $2n + 1$ copies of T cycle through the $2n + 1$ edges of each displacement class without repetitions, showing that $K_{2n+1} = \underbrace{T \oplus T \oplus \ldots \oplus T}_{(2n+1)-\text{ times}}$. $\square$

It is worth mentioning that in the proof of Theorem 5.1, it was never used the fact that T is a tree, therefore if this assumption is dropped, the theorem is still true. Next we show an example.

Example 5.1 Consider the cycle C_4, with the vertex labeling shown in Fig. 5.1. It is easy to check that this labeling is in fact a graceful labeling of C_4. Therefore, according to Theorem 5.1, K_9 is decomposable into 9 copies of C_4. See the rotations that take place in Fig. 5.2.

Fig. 5.1 A vertex labeling of C_4

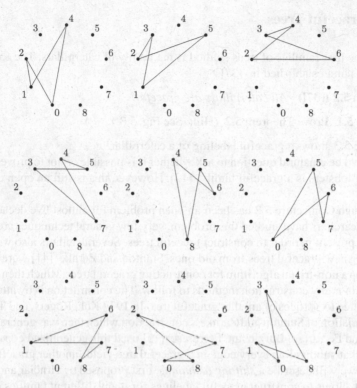

Fig. 5.2 C_4 as a subgraph of K_9, all rotations from a graceful labeling

Due to Theorem 5.1, to characterize the set of graceful trees has become a very well-studied and challenging problem that is still open nowadays. In fact, the following conjecture appeared in [37] and it is known as either the graceful tree conjecture or the Ringel–Kotzig conjecture.

Conjecture 5.2 ([37]) Every tree is graceful.

It is worth to point out that some mathematicians have had their doubts on the hardness involved to find a solution of this conjecture. The following words by Gallian found in [19] will illustrate this fact:

> In 1986 I had two extremely strong undergraduate summer research students named Dog Jungers and Mike Reid whom I wanted to challenge. Both of them were winers in the International Mathematical Olympiad in 1983 and 1984. Of course, I knew that the graceful tree conjecture was considered to be notoriously difficult but I though that these two exceptional problem solvers might be able to come up with a fresh approach to this problem, but were not able to make any substantial progress. This convinced me that the graceful tree conjecture deserved its notoriety for being difficult.

Although the conjecture seems to be very hard to solve, some results on particular families of trees have appeared in the literature. We are going to study some of these results in the next section. For integers $m \leq n$, let $[m, n] = \{m, m + 1, ..., n\}$.

5.2   Graceful Trees

One of the first families of trees studied is the family of caterpillars. The following result was first established in [37].

Theorem 5.2 ([37])   *All caterpillars are graceful.*

Exercise 5.1   Prove Theorem 5.2. (Hint. See Fig. 5.3.)

Figure 5.3 shows a graceful labeling of a caterpillar.

It would be a natural question to ask whether it is possible or not to prove that the family of lobsters is a graceful family [12]. However, this is still an open question so far.

Although Conjecture 5.2 has been an open problem for almost five decades, and many researchers have studied this problem, very few general techniques are known up to the present in order to construct graceful trees. Several authors also worked in producing new graceful trees from old ones. Stanton and Zarnke [41] were the first to develop a non-trivial algorithm for constructing graceful trees, which then became the basis of many construction methods to follow. The construction is by attaching a graceful tree to vertices of another graceful tree. In 1977 Koh, Rogers, and Tan [29] gave a variation of Stanton and Zarnke's construction which then was generalized by Burzio and Ferrarese. Burzio and Ferrarese [15] used this generalized construction to show that subdividing every edge in a graceful tree yields another graceful tree.

Next, we will describe a *shifting technique*, first proposed by Hrnčiar and Haviar [27], that allows to construct graceful labelings for many different families of trees. An example of this fact is the following result.

Theorem 5.3 ([27])   *All trees of diameter at most five are graceful.*

We will show results due to Balbuena et al. [10] that illustrate the use of this shifting technique.

5.2.1   The Shifting Technique

The technique consists in transforming a graceful tree T into a new graceful tree T' of the same size.

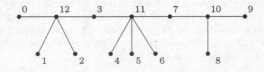

Fig. 5.3   A graceful labeling of a caterpillar

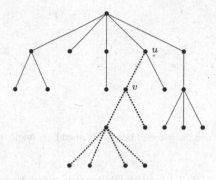

Fig. 5.4 A tree T with the edges of $T_{u,v}$ in *dotted lines*

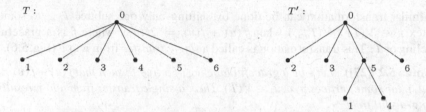

Fig. 5.5 A graceful tree T and the graceful tree T' obtained by a pair transfer from 0 to 5

Definition 5.1 ([27]) Let T be any tree and let uv be any edge of T. We denote by $T_{u,v}$, the subtree of T induced by the set of vertices:

$$V(T_{u,v}) = \{w \in V(T) : \ w = u \text{ or } v \text{ belongs to the unique } uw \text{ path in } T\}.$$

Figure 5.4 shows a tree with a selected edge uv. The dotted edges correspond to the edges of $T_{u,v}$.

Next, we show how to transform a graceful tree T with a graceful labeling f into a new graceful tree T' of equal size. The new tree T' is obtained from T by shifting pairs of subtrees T_{u,u_1}, T_{u,u_2} to a vertex $v \in V(T) \setminus (V(T_{u,u_1}) \cup V(T_{u,u_2}))$. It is requested that $f(u) + f(v) = f(u_1) + f(u_2)$ in order to be able to perform the shift. This transformation was introduced in [27] and called a *pair transfer* from u to v (Fig. 5.5). Roughly speaking, this notation means to identify vertex u with vertex v and to hang from v the two subtrees T_{u,u_1} and T_{u,u_2}. All vertices keep the same labels assigned by f.

Lemma 5.1 ([27]) *Let f be a graceful labeling of a tree T, such that $f(u) + f(v) = f(u_1) + f(u_2)$, for some vertices $u, v, u_1, u_2 \in V(T)$. Then, a pair transfer from u to v results in a graceful tree.*

Proof All edge induced labels of T by f remain the same on the new tree, except for the edge induced labels of the two edges that have been shifted. That is to say, the edges uu_1 and uu_2 that have become edges vu_1 and vu_2. Now, the equality

T :

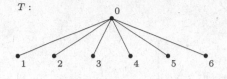
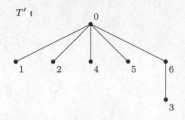

Fig. 5.6 A graceful tree T and the graceful tree T' obtained by a single transfer from 0 to 6

$f(u) + f(v) = f(u_1) + f(u_2)$ implies that $|f(u) - f(u_1)| = |f(v) - f(u_2)|$ and $|f(u) - f(u_2)| = |f(v) - f(u_1)|$. Therefore, the result follows. □

A similar transformation can be done by shifting only one subtree T_{u,u_1} to some vertex $v \in V(T) \setminus V(T_{u,u_1})$ when $f(u) + f(v) = 2f(u_1)$, where f is a graceful labeling of T. This transformation is called a *single transfer* from u to v (Fig. 5.6).

Lemma 5.2 ([27]) *Let f be a graceful labeling of a tree T, such that $f(u) + f(v) = 2f(u_1)$, for some vertices $u, v, u_1 \in V(T)$. Then, a single transfer from u to v results in a graceful tree.*

Exercise 5.2 Prove Lemma 5.2.

Next, we will use these ideas in order to prove that trees having either an even or a quasi even degree sequence, which we introduce next, are graceful. Let T be any tree. If T has even diameter D and it is rooted at its central vertex a, then $V(T) = \cup_{1 \le i \le D/2} L_i \cup \{a\}$, where $L_i = \{v \in V(T) : d_T(v, a) = i\}$. If T has odd diameter D and it is rooted at its two central vertices a, b, then $V(T) = \cup_{1 \le i \le (D-1)/2} L_i \cup \{a, b\}$, where $L_i = \{v \in V(T) : d_T(v, a) = i$ and $d_T(v, b) = i + 1\} \cup \{v \in V(T) : d_T(v, a) = i + 1$ and $d_T(v, b) = i\}$. Following the standard terminology, we will refer to L_i as the levels of the rooted tree.

The following definitions and results were first introduced by Balbuena et al. [10].

Definition 5.2 ([10]) A rooted tree T has an *even degree sequence* if every vertex has even degree except for exactly one root and the leaves. A rooted tree T with diameter D has a *quasi even degree sequence* if every vertex has even degree except for exactly one root, the leaves and the vertices in level $L_{\lfloor D/2 \rfloor - 1}$.

Figures 5.7 and 5.8 illustrate examples of the previous definitions.

Theorem 5.4 ([10]) *Every tree with an even or quasi even degree sequence is graceful.*

In order to prove Theorem 5.4, we introduce the following results. The first one establishes a relation between the size and the diameter of a tree with a (quasi) even degree sequence.

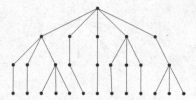

Fig. 5.7 A tree with an even degree sequence and even diameter

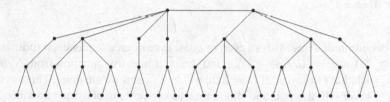

Fig. 5.8 A tree with a quasi even degree sequence and odd diameter

Lemma 5.3 ([10]) *Let T be a tree of size m and even diameter D. If T has an even degree sequence, then $m + D/2$ is even. If T has a quasi even degree sequence, then $m + D/2$ is odd.*

Proof Assume that T has either an even or a quasi even degree sequence. Assume that a is the root of T. Then $d_T(a)$ is odd. This implies that $|L_1|$ is odd. It is clear that, for all $i \in [2, D/2]$, $|L_i| = \sum_{w \in L_{i-1}} (d_T(w) - 1) = \sum_{w \in L_{i-1}} d_T(w) - |L_{i-1}|$, or equivalently, we have

$$|L_i| + |L_{i-1}| = \sum_{w \in L_{i-1}} d_T(w), \ \forall i \in [2, D/2]. \tag{5.1}$$

Since for every $i \in [2, D/2 - 1]$, we have that $d_T(w)$ is even, $w \in L_i$, it follows that $\sum_{w \in L_{i-1}} d_T(w)$ is even. Thus, from (5.1) we get that $|L_i| + |L_{i-1}|$ is even. Hence, $|L_i|$ and $|L_{i-1}|$ have the same parity for every $i \in [2, D/2 - 1]$. That is, since $|L_1|$ is odd, it follows that $|L_1|, |L_2|, \ldots, |L_{D/2-1}|$ are all odd. Therefore, if T has an even degree sequence, then $|L_{D/2}|$ is also odd, and therefore, $m = \sum_{i=1}^{D/2} |L_i|$ and $D/2$ have the same parity.

On the other hand, if T has a quasi even degree sequence, then $|L_{D/2}|$ is clearly even and this implies that the parity of m and the parity of $D/2$ are different. $\square$

Theorem 5.5 ([10]) *Let T be a tree with either an even or a quasi even degree sequence and with even diameter D. Then T is graceful.*

Proof Let T be a tree of size m, with an even or a quasi even degree sequence and with even diameter D. Assume that T has been rooted at its central vertex a and consider the set of $d_T(a)$ labels $X = [0, (d_T(a) - 1)/2] \cup [m - (d_T(a) - 1)/2, m - 1]$. We want to show that any bijective function from $N_T(a)$ to X can be extended to a graceful labeling f of $V(T)$ with the property that $f(a) = m$.

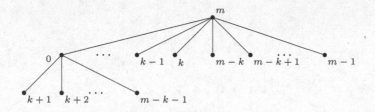

Fig. 5.9 The tree T^0

By definition of a tree with an even or quasi even degree sequence, it follows that $d_T(a) \geq 3$. Let $k = (d_T(a) - 1)/2$ and define a bijective function from $N_T[a]$ to $X \cup \{m\}$ such that $f(a) = m$. If $D = 2$ then $T \cong K_{1,n}$ and we are done. Thus, assume that $D \geq 4$. We will use pair and single transfers in order to construct a sequence of graceful trees, all of them having the same size, that will finish with the original tree

$$T^0, T^{m-1}, T^1, T^{m-2}, \ldots, T, \tag{5.2}$$

where T^0 is the tree that appears in Fig. 5.9.

We will distinguish two cases according to the parity of $d_T(0)$.

Case 1. Assume first that $d_T(0)$ is even. The children of 0 in T^0 form the set $[k + 1, m - k - 1]$. That is, except for the number $m - k - 1$, the rest of the elements can be associated in pairs whose sum is $m - 1$. For instance, this pairing can be performed as follows: $(k + 1) + (m - k - 2) = (k + 2) + (m - k - 3) = \ldots$. Let $k_0 = d_T(0)/2$. The graceful tree T^{m-1} is obtained from T^0 by successive transfers from 0 to $m-1$ shifting the pairs of subtrees of T^0, $T^0_{0,k+k_0}$ and $T^0_{0,m-1-k-k_0}$, $T^0_{0,k+k_0+1}$ and $T^0_{0,m-2-k-k_0}$, $\ldots$. Thus, in the new tree T^{m-1} the children of $m - 1$ are labeled with the set of consecutive labels: $[k + k_0, m - 1 - k - k_0]$ and 0 is still adjacent to: $[k + 1, k + k_0 - 1] \cup [m - (k + k_0), m - k - 1]$. Hence, $d_{T^{m-1}}(0) = d_T(0)$. Next, we describe how to obtain T^1 from T^{m-1}. We proceed in a similar way. Except for $k + k_0$, all the elements in the list of children of $m - 1$ in T^{m-1} can be associated in pairs whose sum is m. For instance, $(k + k_0 + 1) + (m - 1 - k - k_0) = (k + k_0 + 2) + (m - 2 - k - k_0) = \ldots$. Let $k_{m-1} = d_T(m - 1)/2$. The graceful tree T^1 is obtained from T^{m-1} by successive transfers from $m - 1$ to 1, shifting the pairs of subtrees of $T^{m-1}_{m-1,k+k_0+k_{m-1}}$ and $T^{m-1}_{m-1,m-k-k_0-k_{m-1}}$, $\ldots$. This implies that the remaining children of $m - 1$ in T^1 are the elements in the set: $[k + k_0, k + k_0 + k_{m-1} - 1] \cup [m - (k + k_0 + k_{m-1} - 1), m - (k + k_0) - 1]$. Thus, $d_{T^1}(0) = d_T(0)$ and $d_{T^1}(m - 1) = d_T(m - 1)$. We continue this procedure inductively by creating the new element T^j in (5.2) from its preceding element T^i. Using pairs of transfers or single transfers from i to j, in such a way that the following conditions hold:

– The children of j in T^j form a set of consecutive labels.
– $d_{T^j}(i) = d_T(i)$.
– The children of i are labeled in T as they have been labeled in T^j.

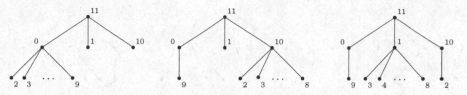

Fig. 5.10 First steps to label a tree with even degree sequence and even diameter

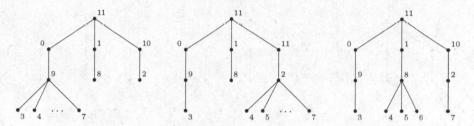

Fig. 5.11 Final steps to label a tree with even degree sequence and even diameter

Assume that T has a quasi even degree sequence. Otherwise, if T has an even degree sequence, then this procedure ends up when all the vertices of T have been labeled. Figures 5.10 and 5.11 illustrate this procedure.

The procedure explained so far ends up when T^{i_0} in the sequence (5.2) has $d_T(i_0)$ odd. This implies, in particular, that $i_0 \in L_{D/2-1}$ in T. The children of i_0 in T^{i_0} form a consecutive set of labels. Let T^{j_0} be the successor tree of T^{i_0} in (5.2). By Lemma 5.3, if m is even then $D/2 - 1$ is also even, which implies that $i_0 + j_0 = m$, since in this case all the levels L_j with j even start with a transfer from u to v such that $u + v = m$. Similarly, by Lemma 5.3, if m is odd, then $D/2 - 1$ is also odd. Hence, $i_0 + j_0 = m - 1$, since in this case all the levels L_j with j odd start with a transfer from u to v such that $u + v = m - 1$. Therefore, we construct T^{j_0} from T^{i_0} by a successive pair or single transfers from i_0 to j_0 such that i_0 remains adjacent to its smallest child, if m is even, and remains adjacent to its largest child otherwise. The vertex i_0 remains adjacent to the vertex $\lfloor m/2 \rfloor = (i_0 + j_0)/2$ and to all pairs of vertices that are necessary in order to complete its odd degree. After this transformation the children of j_0 in T^{j_0} form two disjoint sets of consecutive labels. Also, $d_{T^{j_0}}(i_0) = d_T(i_0)$ and now the children of i_0 are labeled in T as they have been labeled in T^{j_0}. We continue in this way by constructing T^j from the preceding T^i in (5.2), following this new procedure.

- The children of j in T^j form two disjoint sets of consecutive labels.
- $d_{T^j}(i) = d_T(i)$.
- The children of i are labeled in T as they have been labeled in T^j.

Figures 5.12 and 5.13 show this procedure.

Case 2. Assume now that $d_T(0)$ is odd. In this case, all vertices in the same level are forced to have odd degree. Thus, T has a quasi even degree sequence

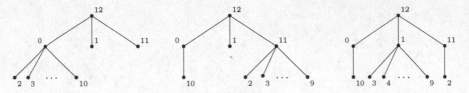

Fig. 5.12 First steps to label a tree with quasi even degree sequence and even diameter

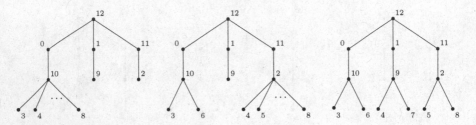

Fig. 5.13 Final steps to label a tree with quasi even degree sequence and even diameter

with diameter 4. Hence, we can apply the last algorithm described in Case 1, and therefore, we are done.   □

Next, we state and prove the following theorem.

Theorem 5.6 ([10]) *Let T be a tree with either an even or a quasi even degree sequence and with odd diameter D. Then T is graceful.*

Proof Let T be a tree with odd diameter D and let a and b be the two adjacent central vertices of T. If T has either an even or quasi even degree sequence, then either a has odd degree and b even degree or vice versa. Without loss of generality assume that $d_T(a)$ is odd.

Let us consider the tree T' obtained from T by identifying the two vertices a and b. Then $d_{T'}(a) = d_T(a) + d_T(b) - 2$ is odd. Also, the diameter of T' is equal to $D - 1$, which is even, and the rest of the degrees of T' remain the same as they were in T. Thus, T' has either an even or a quasi even degree sequence. Hence, applying Theorem 5.5, we can obtain a graceful labeling f of T' with the following two properties:

- $f(a) = m - 1$, where $m = |E(T)|$ and
- $f(N_{T'}(a)) = [0, (d_{T'}(a) - 1)/2] \cup [m - 1 - (d_{T'}(a) - 1)/2, m - 2]$.

In particular, $f(N_T(a) \setminus \{b\}) \subset \{0, m-2, 1, m-3, \ldots\}$. Thus, the labels of $N_T(a) \setminus \{b\}$ can be paired up in such a way that the sum of the labels forming a pair is $m - 2$.

At this point, let T'' be a tree obtained from T' by introducing a new vertex a' and joining it with the root a of T'. Then $d_{T''}(a) = d_{T'}(a) + 1$ and hence is even. Also $|V(T'')| = |V(T)|$. Let us define a new labeling g on T'' as follows: $g(a') = 0$ and $g(v) = f(v) + 1$, for all $v \neq a'$. Clearly, g is a graceful labeling of T'' and the labels of $N_{T''}(a)$ are the elements on the set $X = [0, d_{T''}(a)/2] \cup [m - d_{T''}(a)/2, m - 1]$.

Thus, the labels of $N_T(a) \setminus \{b\}$ can be paired up so that the sum of each pair is m. Hence, we obtain a graceful labeling of T from the labeling g of T'' by pairs transfers from a to a', shifting $(d_T(a)-1)/2$ pairs of subtrees T''_{a,u_i}, T''_{a,u_j}, with $u_i, u_j \in N_T(a) \setminus \{b\}$, such that $g(a_i) + g(a_j) = m$. Therefore, we are done. $\square$

Exercise 5.3 Show that every tree is a subtree of a graceful tree. (Hint: use the graceful tree obtained in Theorem 5.5.)

The previous exercise motivates the following open problem.

Problem 5.1 For a given tree T, find any one tree of the smallest graceful trees, namely tree T', such that T is a subtree of T'.

Notice that to answer this problem immediately solves the graceful tree conjecture.

5.2.2 Applications of Shifting Techniques

Next, we apply the shifting technique in a different way to construct graceful labelings for path-like trees and T_p-trees.

5.2.2.1 Path-Like Trees

Path-like trees were introduced by Barrientos in his Ph.D thesis [11]. We embed, without crosses, the path P_n as a subgraph of the two dimensional grid (that is to say, the graph $P_m \times P_l$ where m and l are at least n). Given such an embedding, we consider the ordered set of subpaths $L_1, L_2, \ldots, L_k$ which are maximal straight segments in the embedding, where the end of L_i is the beginning of L_{i+1}, for any $i \in [2, k-1]$. Let

1. $L_i \cong P_2$ for some $i \in [2, k-1]$.
2. $V(L_i) = \{u_0, v_0\}$ where $\{u_0\} = V(L_{i-1}) \cap V(L_i)$ and $\{v_0\} = V(L_i) \cap V(L_{i+1})$.

Suppose that a vertex $u \in V(L_{i-1})$ and a vertex $v \in V(L_{i+1})$ are at distance 1 in the grid and that

$$d_{L_{i-1}}(u, u_0) = d_{L_{i+1}}(v, v_0).$$

Definition 5.3 ([11]) An *elementary transformation* of the path consists in replacing the edge u_0v_0 by the new edge uv.

Definition 5.4 ([11]) A tree of order n is a *path-like tree* when it can be obtained after a set of elementary transformations on an embedding of P_n in the two dimensional grid.

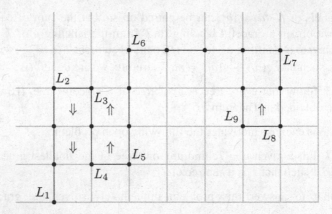

Fig. 5.14 An embedding of a path with its possible elementary transformations indicated

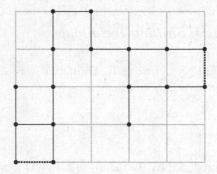

Fig. 5.15 A path-like tree

Figure 5.14 shows an embedding of a path in the grid, with all L_i's specified and with all possible elementary transformations indicated. Figure 5.15 shows a path-like tree in the grid. The dotted segments indicate the edges that were originally in the embedding of the path and that have been replaced by new edges to obtain the path-like tree.

Exercise 5.4 Show that all path-like trees are graceful. (*Hint*: Since every path is a caterpillar, we can consider the graceful labeling suggested in Exercise 5.1.)

Although every path-like tree is graceful, it is not always easy to determine when a given tree is a path-like tree. In fact, Bača et al. [8] proposed the following problem which is still open.

Problem 5.2 Determine the complexity of deciding if a given tree of maximum degree at most 4 is a path-like tree.

5.2.2.2 T_p-Trees

Next, we discuss the concept of T_p-trees, which is, in some sense, the opposite of path-like trees, since instead of obtaining "the tree from the path," we obtain "the path from the tree." T_p-trees were introduced by Hegde and Shetty in [25].

Definition 5.5 ([25]) Let T be a tree and u_0, v_0 be two adjacent vertices in T. Let there be two leaves $u, v \in V(T)$ such that the length of the uu_0 path and the length of the vv_0 path are equal, then the edge u_0v_0 is called a *transformable edge*. Consider the transformation of T in which the edge u_0v_0 is deleted and the vertices u, v are joined by an edge uv. Such a transformation is called an *elementary parallel transformation*.

Definition 5.6 ([25]) If by a sequence of elementary parallel transformations a tree T can be reduced to a path, then T is called a T_p-*tree* and any such sequence is called a *parallel transformation* of T.

The tree T showed in Fig. 5.16 is a T_p-tree. Figure 5.17 shows all the steps that allow us to transform T into a path of the same size.

Exercise 5.5 Show that the family of T_p-trees is a family of graceful trees [25].

One of the questions that comes to mind when comparing the families of path-like trees and of T_p-trees is to determine the relation existing between them. It is clear that there are trees which are path like-trees and T_p-trees at the same time. The relation existing among path-like trees and T_p-trees is as shown in the Venn diagram that appears in Fig. 5.18 [33]. To verify this diagram we propose the next two exercises.

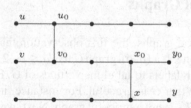

Fig. 5.16 A T_p-tree

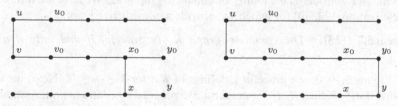

Fig. 5.17 A transformation of a T_p-tree into a path

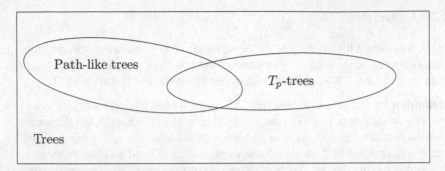

Fig. 5.18 The relation existing among path-like trees and T_p-trees

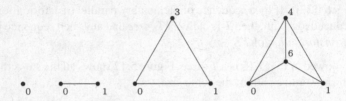

Fig. 5.19 Graceful labelings of K_n, for $1 \leq n \leq 4$

Exercise 5.6 Find a path-like tree which is not a T_p-tree.

Exercise 5.7 Find a T_p-tree which is not a path-like tree.

5.3 Other Graceful Graphs

When dealing with general graphs, the first observation that should be taken into account is the following one. If a (p, q)-graph G has $q \leq p-2$, then G is not graceful, since there are not enough labels to label the vertices of G. Even if $q \geq p-1$, it is not possible to guarantee that G is graceful. For instance, it is an easy exercise to show that the graph $C_3 \cup K_2$ is not a graceful graph. Next, we study the gracefulness of certain classes of well-studied graphs which are not trees. The first family of graphs that we consider is the family of complete graphs K_n. We have the following characterization [23, 39] for which we provide a constructive proof.

Theorem 5.7 ([23])   *The complete graph K_n is graceful if and only if $n \in \{1, 2, 3, 4\}$.*

Proof Figure 5.19 shows graceful labelings of K_n, for $1 \leq n \leq 4$. Next, we will show that if $n \geq 5$, then K_n is not graceful. We proceed by contradiction. Assume to the contrary that there is $n \geq 5$ for which the complete graph K_n is graceful. Since $|E(K_n)| = \binom{n}{2}$, it follows that the vertices of K_n have to be labeled using n elements from the set $[0, \binom{n}{2}]$, by any graceful labeling f of K_n. Also, the set of induced edge labels must be the set $[1, \binom{n}{2}]$.

First of all, notice that the only way to get the edge label $\binom{n}{2}$ is by using the pair of vertex labels 0 and $\binom{n}{2}$. Next, we will create the edge label $\binom{n}{2} - 1$. We have two possibilities in order to do this. We can create this edge label either using the pair of vertices 0 and $\binom{n}{2} - 1$, or the pair of vertices 1 and $\binom{n}{2}$. In both cases, we will create edge induced labels $\binom{n}{2}$, $\binom{n}{2} - 1$ and 1.

Case 1. First of all assume that we have used the vertex labels $0, \binom{n}{2}$ and $\binom{n}{2} - 1$. This configuration forbids the labels 1 and $\binom{n}{2} - 2$, since otherwise, an induced edge label appears twice. Thus, in order to create the edge label $\binom{n}{2} - 2$, we have to use the pair $\{2, \binom{n}{2}\}$. That is, the vertex labels used so far are $0, 2, \binom{n}{2} - 1$ and $\binom{n}{2}$. Now, suppose we want to construct the edge label $\binom{n}{2} - 4$. We cannot use as vertex labels any of the elements in the set $\{3, 4, \binom{n}{2} - 3, \binom{n}{2} - 2\}$, otherwise, the edge label 1 or 2 appears twice. Thus, the only possibility is to use $\{0, \binom{n}{2} - 4\}$. See Fig. 5.20.

Hence, we get that the vertex labels used so far are $0, 2, \binom{n}{2} - 4, \binom{n}{2} - 1$ and $\binom{n}{2}$. Furthermore, the edge induced labels obtained so far include the set $\{1, 2, 3, 4, \binom{n}{2} - 6, \binom{n}{2} - 4, \binom{n}{2} - 3, \binom{n}{2} - 2, \binom{n}{2} - 1, \binom{n}{2}\}$. Now, our goal is to obtain the edge label $\binom{n}{2} - 5$. We can do this using anyone of the following pairs of vertex labels $\{0, \binom{n}{2} - 5\}, \{1, \binom{n}{2} - 4\}, \{2, \binom{n}{2} - 3\}, \{3, \binom{n}{2} - 2\}, \{4, \binom{n}{2} - 1\}$, and $\{5, \binom{n}{2}\}$. We will see that, independently of the pair that we use, we always reach a contradiction. Let us start using pair $\{0, \binom{n}{2} - 5\}$. Then, we get the edge label 1 repeated, since it appears as induced edge label of the edges with end vertices $\{\binom{n}{2} - 4, \binom{n}{2} - 5\}$

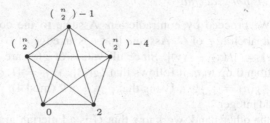

Fig. 5.20 Some induced edge labels of K_n

and $\{\binom{n}{2} - 1, \binom{n}{2}\}$. Next, we use the pair $\{1, \binom{n}{2} - 4\}$. Again, we get the edge label 1 repeated, a contradiction. Recall that a previous reasoning forbids $\{3, 4, \binom{n}{2} - 5, \binom{n}{2} - 3, \binom{n}{2} - 2\}$, as possible vertex labels. Thus, the only pair that remains is $\{5, \binom{n}{2}\}$. Notice that in this case, we get the edge label 3 repeated, a contradiction. This shows that following the first option we do not get any graceful labeling of K_n, for $n \geq 6$. For $n = 5$, in the labeling we got, there are two edges labeled 4. These are the edges labeled 4 and $\binom{n}{2} - 6$. Therefore, we have not been able to obtain any graceful labeling of K_n, for $n \geq 5$.

Case 2. Similar calculations show that if we use the vertex labels $0, 1, \binom{n}{2}$ to begin with, we obtain a similar result. Therefore, K_n is not graceful, for $n \geq 5$.   □

Another well-studied family of graphs for which it is easy to obtain a graceful labeling is the family of complete bipartite graphs.

Theorem 5.8 ([23, 37]) *Let m and n be positive integers. Then, the complete bipartite graph $K_{m,n}$ is graceful.*

Proof Assume that $m \leq n$ and let X and Y be the stable sets of $V(K_{m,n})$, with $|X| = m$. Consider any labeling $f : V(K_{m,n}) \rightarrow [0, m - 1] \cup \{im\}_{i=1}^{n}$ such that $f(X) = [0, m - 1]$ and $f(Y) = \{im\}_{i=1}^{n}$. An easy check shows that f is a graceful labeling of $K_{m,n}$.   □

Although it is relatively easy to prove that all complete bipartite graphs are graceful, only small cases of complete multipartite graphs are known to be graceful. For instance, the graphs $K_{1,m,n}$ [5] and $K_{1,1,m,n}$ [22]. Thus, we propose the following research problem.

Problem 5.3 Characterize complete multipartite graphs that are graceful.

Another interesting family that has been wildly considered in the world of graphs labelings is the family of cycles. Hence, it is a natural question to ask if we can characterize the set of graceful cycles. However, in order to do this, it is convenient to introduce first the following lemma concerning Eulerian graphs.

Lemma 5.4 ([37]) *Let G be a Eulerian graph of size q, where $q \equiv 1$ or 2 (mod 4). Then, G is not graceful.*

Proof We proceed by contradiction. Assume to the contrary that there exists a graceful labeling f of G. Assume that e is an edge of G with end vertices x and y. Let $g(e) = |f(x) - f(y)|$. Since all values of $g(e)$ are different, and we use the labels from 0 up to q, it follows that $\{g(e)\}_{e \in E(G)} = [1, q]$. In particular, we obtain $\sum_{e \in E(G)} g(e) = \sum_{i=1}^{q} i$. Using that $q \equiv 1$ or 2 (mod 4), we get that the sum $\sum_{i=1}^{q} i$ is an odd integer.

On the other hand, we know that G is a Eulerian graph. Thus, $G = \oplus_{i=1}^{k} C_{n_k}$, $n_k \geq 3$ (see Exercise 1.7). Let us consider any arbitrary cycle in this decomposition, namely C_{n_i}, and let $\{g(e) \; : \; e \in E(C_{n_i})\}$ be the edge labels on the edges of C_{n_i}, induced by the graceful labeling f. Then $\sum_{e \in E(C_{n_i})} g(e)$ is even, since in this sum every vertex label $f(x)$ is repeated twice (either adding or subtracting). Hence, we can add all these extended sums over all cycles in the decomposition, and we get that $\sum_{e \in E(G)} g(e)$ is even. Therefore, the desired contradiction has been reached.

$\square$

By Lemma 5.4, it is clear that if a cycle C_n is graceful, then $n \equiv 0$ or $3 \pmod 4$. It turns out that this necessary condition is also sufficient.

Theorem 5.9 ([37]) *The cycle C_n is graceful if and only if $n \equiv 0$ or $3 \pmod 4$.*

Proof The only thing that remains to be proven is that if $n \equiv 0$ or $3 \pmod 4$ then C_n is graceful. Let $V(C_n) = \{v_i\}_{i=1}^n$ and $E(C_n) = \{v_i v_{i+1}\}_{i=1}^{n-1} \cup \{v_1 v_n\}$. We distinguish two cases.

Assume first that $n \equiv 0 \pmod 4$. We define a labeling $f : V(C_n) \to [0, n]$ as follows:

$$f(v_i) = \begin{cases} (i-1)/2, & \text{if } i \text{ is odd}, \\ n+1-i/2, & \text{if } i \text{ is even and } i \leq n/2, \\ n-i/2, & \text{if } i \text{ is even and } i > n/2. \end{cases}$$

An easy check shows that f is a graceful labeling of C_n.

Assume now that $n \equiv 3 \pmod 4$. We define a labeling $f : V(C_n) \to [0, n]$ as follows:

$$f(v_i) = \begin{cases} n+1-i/2, & \text{if } i \text{ is even}, \\ (i-1)/2, & \text{if } i \text{ is odd and } i \leq (n-1)/2, \\ (i+1)/2, & \text{if } i \text{ is odd and } i > (n-1)/2. \end{cases}$$

An easy check shows that f is a graceful labeling of C_n.

$\square$

Figure 5.21 shows graceful labelings of the cycles C_7 and C_8.

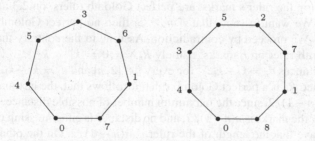

Fig. 5.21 Graceful labelings of the cycles C_7 and C_8

5.4   Golomb Rulers

A concept that is closely related to graceful labelings, and in particular to graceful labelings of complete graphs, is the one of *Golomb ruler* (see [7, 21, 38]), named after Solomon Golomb. Golomb rulers have applications to different problems that take place in real life. For instance, the following problem described by Gardner in [21] has a direct connection with Golomb rulers. Assume that a radio astronomer needs to place satellite dishes along a straight line in such a way that every possible distance $1, 2, \ldots, N$ is achieved between some pair of them at most once. It is reasonable to use the minimum number of satellites in order to minimize cost. Next, we define Golomb rulers and the relationship existing between the problem introduced by Gardner and Golomb rulers will be evident. *A Golomb ruler* is a straight edge containing labeled marks such that the distance between any two marks is distinct. The 0 mark and the last mark are the ends of the ruler, so that the ruler has n marks. The length of the ruler is the largest mark.

Definition 5.7 A Golomb ruler that is of minimum length for a given number of marks is called an *optimal Golomb ruler*. A *perfect Golomb ruler* of n marks is a Golomb ruler in which every integer from 1 up to the length of the ruler can be measured as the distance of exactly two marks.

From our discussion so far, it should be clear that the complete graph K_n is graceful if and only if there exists a perfect Golomb ruler on n marks. The reason is that if there is a perfect Golomb ruler on n marks then we can use these marks to gracefully label the vertices of K_n. If K_n is graceful, then the labels of the vertices of K_n can be used as marks on any Golomb ruler on n marks. Thus, we obtain the following theorem as a direct consequence of Theorem 5.7. However, we propose another proof to show a different approach.

Theorem 5.10 ([39]) *A perfect Golomb ruler of n marks exists if and only if $n \in \{1, 2, 3, 4\}$.*

Proof A ruler with mark 0 is trivially a perfect Golomb ruler. Next, the rulers $R_2 = \{0, 1\}$, $R_3 = \{0, 1, 3\}$ and $R_4 = \{0, 1, 4, 6\}$ where the elements of the sets are representing the ruler's marks, are perfect Golomb rulers, on 2,3 and 4 marks, respectively. We want to show that if $n \geq 5$, then no perfect Golomb ruler on n marks exists. We proceed by contradiction. Assume to the contrary that there is a perfect Golomb ruler on n marks, namely $R_n = \{0 = k_1 < k_2 < \ldots < k_n\}$. We define the distances $d_i = k_i - k_{i-1}$, for every $i \in [2, n]$ and $d'_i = k_i - k_{i-2}$, for every $i \in [3, n]$. Since R_n is a perfect Golomb ruler, it follows that, the last mark is placed at position $n(n-1)/2$, since the maximum number of possible distances that we can obtain among the marks is $n(n-1)/2$, and no distance is either missing or repeating. Hence, we have that the length of the ruler is $n(n-1)/2$. On the other hand, it is clear that the length of the ruler is given by the sum

Fig. 5.22 A possible configuration of R_n

Fig. 5.23 A possible configuration of R_n

$$\sum_{i=2}^{n} d_i = n(n-1)/2. \tag{5.3}$$

Using (5.3) and the fact that all the $n-1$ distances $\{d_i\}_{i=2}^{n}$ are distinct, we obtain $\{d_i\}_{i=2}^{n} = \{i\}_{i=1}^{n-1}$.

Next, we place these distances on R_n. Let us start by placing distance 1 on R_n. Assume that distance 1 and distance $n-1$ are not next to each other in R_n. Then, there is a distance l, $l \neq n-1$ such that l and 1 are next to each other, in the ruler. Then, there is some element in $\{d_i'\}_{i=3}^{n}$ equal to $l+1 \leq n-1$, which is impossible since $\{d_i'\}_{i=3}^{n} \cap \{d_i\}_{i=2}^{n} = \emptyset$. Thus, the distances 1 and $n-1$, both in $\{d_i\}_{i=2}^{n}$ are next to each other in R_n. Furthermore, 1 has to be either the first end distance or the last end distance in the ruler. Otherwise, it would be another distance next to it, and by using a similar reasoning as above, we will reach to a contradiction. Hence, without loss of generality, we can assume the configuration that appears in Fig. 5.22 taking place in R_n. Observe now that, unless $n=3$, distance 2 cannot be formed by any mark next to distance 1. Again distance 2 cannot be next in the ruler to any distance strictly smaller than $n-2$, by a similar reasoning as before. Hence distance 2, again using a similar reasoning as before, must be the distance belonging to the set $\{d_i\}_{i=2}^{n}$, located at the far right end of R_n. Thus, at this point, we have that R_n has the configuration that appears in Fig. 5.23. Following the same reasoning, if we try to include any other element of $\{d_i\}_{i=2}^{n}$, will lead us to conclude that such distance must appear in one end of the ruler. This is a contradiction since there are no any other free ends of the ruler left. Therefore, no R_n exists for $n \geq 5$. $\qquad\square$

The study of Golomb rulers is still a very active area of research and very few optimal Golomb rulers are known. Table 5.1 shows the optimal Golomb rulers known so far. For big values of n, some of these Golomb rulers have been obtained using very heavy computer calculations.

It is also interesting to observe that Golomb rulers are also an alternative way to understand Sidon sets. The following indirect argument shows that Sidon sets and Golomb rulers are equivalent concepts. Assume to the contrary that S is a Sidon set which is not a Golomb ruler. Since S is not a Golomb ruler, it follows that there exists a subset $\{a_1, a_2, a_3, a_4\} \subset S$ of cardinality at least 3 such that, $a_1 - a_2 = a_3 - a_4$, that is, $a_1 + a_4 = a_2 + a_3$. Thus, S is not a Sidon set, a contradiction.

Using a similar argument, we can show that all Golomb rulers are Sidon sets. Finally, we will comment the crucial role that Golomb rulers had in order to solve

Table 5.1 Golomb rulers [28]

Order	Length	Marks	Found	In	Proved	By
1	0	0				
2	1	0 1				
3	3	0 1 3				
4	6	0 1 4 6				
5	11	0 1 4 9 11	1952	[7]	1967?	Hand search
		0 2 7 8 11				
6	17	0 1 4 10 12 17	1952	[7]	1967?	Hand search
		0 1 4 10 15 17				
		0 1 8 11 13 17				
		0 1 8 12 14 17				
7	25	0 1 4 10 18 23 25	1952	[7]	1967?	Hand search
		0 1 7 11 20 23 25				
		0 1 11 16 19 23 25				
		0 2 3 10 16 21 25				
		0 2 7 13 21 22 25				
8	34	0 1 4 9 15 22 32 34	1952	[7]	1972	Hand search
9	44	0 1 5 12 25 27 35 41 44	1972	[32]	1972	Computer search
10	55	0 1 6 10 23 26 34 41 53 55	1967	[36]	1972	Proj. plane const. $p = 9$
11	72	0 1 4 13 28 33 47 54 64 70 72	1967	[36]	1972	Proj. plane const. $p = 11$
		0 1 9 19 24 31 52 56 58 69 72				

Reprint Courtesy of International Business Machines Corporation, ©International Business Machines Corporation, James B. Shearer

one mystery of mathematics that has been around for almost 40 years. We refer to Picard's theorem. In his paper published in 1937, [34] S. Piccard stated and proved the following result.

Theorem 5.11 ([34]) *If X and Y are finite sets of integers whose sets $D(X) = \{x_i - x_j : x_i, x_j \in X, i \neq j$ and $x_i \geq x_j\}$ and $D(Y)$, defined in a similar way, are equal and no $D(X)$ nor $D(Y)$ contain any repeated integer, then $X = \pm Y + c$, where the set $\pm Y + c$ is obtained by adding a constant $c \in \mathbb{Z}$ to every element of Y, or by changing the signs of all the elements of Y and then adding the constant c.*

The problem came up when after the paper was published, nobody was able to understand the argument of the proof. On the other hand, no one was able to find a counterexample neither. While studying Golomb rulers, professor Bloom was confronted with this problem of either understanding the proof of Theorem 5.11 or finding a counterexample for it. After spending some time trying to understand the proof, he decided to change the strategy and go for a counterexample, and to his surprise, he was able to find it. Next, we show the counterexample provided in his paper [13] published in 1977.

Example 5.2 ([13]) Let $X = \{0, 1, 4, 10, 12, 17\}$ and $Y = \{0, 1, 8, 11, 13, 17\}$. Then, $D(X) = D(Y) = [1, 13] \cup \{16, 17\}$.

Hence, this was the end of a theorem that after 38 years, it was shown not to be a theorem. As professor Bloom said in [13], and we code it, we have that, "the sets used in the counterexample can be viewed as the positions of marks on two of the shortest six mark rulers for which all $\binom{n}{2} = 15$ possible measurements are distinct". We learned about Piccard's theorem and its relation with Golomb rulers and Bloom's counterexample during a research visit of Gary Bloom to Barcelona from personal conversations with him in 2007.

5.5 α-Labelings and the Weak Tensor Product

A way to study graceful labelings is using other labelings that are more restrictive than them. For instance, we can use α-labelings for bipartite graphs.

In order to proceed, we define the following product for bipartite graphs that was originally introduced in [40], under the name of weak tensor product.

Definition 5.8 ([40]) Given two bipartite graphs, G_1 and G_2 with stable sets $V(G_1) = X_1 \cup Y_1$ and $V(G_2) = X_2 \cup Y_2$, the *weak tensor product* of G_1 and G_2, denoted by $G_1 \bar\otimes G_2$, is the bipartite graph with vertex set $V(G_1 \bar\otimes G_2) = (X_1 \times X_2) \cup (Y_1 \times Y_2)$ and with $(x_1, x_2)(y_1, y_2) \in E(G_1 \bar\otimes G_2)$ if and only if $x_1 y_1 \in E(G_1)$ and $x_2 y_2 \in E(G_2)$.

Next, we show an example of how the weak tensor product behaves.

Example 5.3 Let G_1 be the cycle of length 4 defined by $V(G_1) = \{0, 1\} \cup \{2, 4\}$ and $E(G_1) = \{02, 12, 14, 04\}$. Similarly, let G_2 the path of length 3 defined by $V(G_2) = \{0, 1\} \cup \{2, 3\}$ and $E(G_2) = \{12, 02, 03\}$. Notice that, for convenience, in this example all the vertices of the graphs G_1 and G_2 are named after the labels assigned by α-labelings of G_1 and G_2. Then, the bipartite graph $G_1 \bar\otimes G_2$ appears in Fig. 5.24.

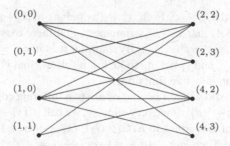

Fig. 5.24 The weak tensor product of G_1 and G_2

Now, let G be a graph with an α-labeling. Following the notation found in [40], we write $V(G) = H \cup L$, to indicate the set of vertices with high and low labels. That is, if k is the characteristic of f, then $f(x) > k$, for every $x \in H$ and $f(y) \leq k$, for every $y \in L$. The next result allows us to create α-labelings (and hence, graceful labelings) of the weak tensor product of two graphs which admit α-labelings.

Theorem 5.12 ([40]) *Let G_1 and G_2 be two bipartite graphs that admit an α-labeling with $V(G_1) = H_1 \cup L_1$ and $V(G_2) = H_2 \cup L_2$, respectively. Then, the graph $G_1 \bar{\otimes} G_2$ also admits an α-labeling.*

Proof Let $q_i = |E(G_i)|$ and let f_i be an α-labeling for G_i, with characteristic k_i, for $i = 1, 2$. Through this proof, whenever we make a reference to an α-labeling for G_i, we will assume we are referring to f_i, and also that $f_i(x) > k_i$, for all $x \in H_i$, $i = 1, 2$. Furthermore, from now on, we will assume that each vertex of G_1 and G_2 receives the name of the label that has assigned to it.

We want to show that the function f' from $V(G_1 \bar{\otimes} G_2)$ to $S \subset \mathbb{Z}$, defined by the rule

$$f'(x_1, x_2) = q_1 x_2 - (q_1 - x_1), \quad (x_1, x_2) \in H_1 \times H_2,$$

$$f'(y_1, y_2) = q_1 y_2 + y_1, \quad (y_1, y_2) \in L_1 \times L_2,$$

is an α-labeling of $G_1 \bar{\otimes} G_2$. Observe that, for each $v \in V(G_1 \bar{\otimes} G_2)$, we obtain $0 \leq f'(v) \leq q_1 q_2$. The lower bound is attained when we evaluate f' at vertex $(0, 0)$. The upper bound when we evaluate f' at vertex (q_1, q_2). Hence, we can see that $f'(V(G_1 \bar{\otimes} G_2)) \subset [0, q_1 q_2]$. Moreover, if $(x_1, x_2) \in H_1 \times H_2$ then,

$$f'(x_1, x_2) = q_1 x_2 - (q_1 - x_1) \geq q_1(k_2 + 1) - (q_1 - k_1 - 1)$$

$$= q_1 k_2 + k_1 + 1 > q_1 k_2 + k_1.$$

Clearly, if $(y_1, y_2) \in L_1 \times L_2$ then $f'(y_1, y_2) \leq q_1 k_2 + k_1$.

Now, we will show that f' is an injective function. Assume that there exist $(x_1, x_2), (x_1', x_2') \in X_1 \times X_2$ such that $f'(x_1, x_2) = f'(x_1', x_2')$. That is, $q_1 x_2 - (q_1 - x_1) = q_1 x_2' - (q_1 - x_1')$, which is equivalent to,

$$q_1(x_2 - x_2') = (x_1' - x_1). \tag{5.4}$$

Then, $x_1' - x_1 \equiv 0 \pmod{q_1}$. Since both, x_1' and x_1 belong to $[k_1 + 1, q_1]$, it follows that $x_1' = x_1$, and using (5.4) that $x_2' x_2$. Assume now, that there exist $(y_1, y_2), (y_1', y_2') \in L_1 \times L_2$ such that $f'(y_1, y_2) = f'(y_1', y_2')$. By a similar reasoning, we obtain $y_1' - y_1 \equiv 0 \pmod{q_1}$. Since both, y_1' and y_1 belong to $[0, k_1]$, it follows that $y_1' = y_1$, and $y_2' = y_2$. This shows that f' is an injective function.

Finally, we want to show that if ab and cd are two different edges in $G_1 \bar{\otimes} G_2$, then $|f'(a) - f'(b)| \neq |f'(c) - f'(d)|$. For every edge xy, we call $|f'(x) - f'(y)|$ the *valence* of the edge. Suppose that two edges in $E(G_1 \bar{\otimes} G_2)$, $(x_1, x_2)(y_1, y_2)$ and $(x_1', x_2')(y_1', y_2')$ have the same valence. Then, $q_1 x_2 - (q_1 - x_1) - q_1 y_2 - y_1 = q_1 x_2' - (q_1 - x_1') - q_1 y_2' - y_1'$, which is equivalent to

$$q_1(x_2 - y_2) + x_1 - y_1 = q_1(x_2' - y_2') + x_1' - y_1'. \tag{5.5}$$

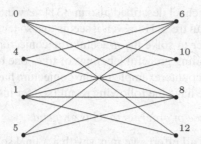

Fig. 5.25 A graceful labeling of the weak tensor product of G_1 and G_2, introduced in Example 5.3

Thus, we obtain, $x_1 - y_1 \equiv x_1' - y_1' \pmod{q_1}$. Now, since x_1y_1 and $x_1'y_1'$ are edges of G_1, we have $x_1 - y_1, x_1' - y_1' \in [1, q_1]$. Hence and by definition of a graceful labeling, it follows that $x_1 - y_1 = x_1' - y'$ and considering (5.5), we get that $x_2 - y_2 = y_2' - x_2'$. But, $x_2y_2, x_2'y_2' \in E(G_2)$. Again, by definition of a graceful labeling, the same valence cannot occur on two distinct edges. Therefore, we conclude that $x_2 = x_2'$ and $y_2 = y_2'$. This proves that f' is a α-labeling of $G_1 \bar{\otimes} G_2$. □

Figure 5.25 shows how the ordered pairs obtained in Fig. 5.24 are transformed into the labels of an α-labeling for the graph $G_1 \bar{\otimes} G_2$.

We close this section by observing that this product is a powerful technique in order to obtain new bipartite graphs which admit α-labelings, and hence, graceful labelings. In fact, El-Zanati et al. [17] have also applied this product in order to obtain *near α-labelings* (also known as *gracious labelings* [24]), a kind of labeling that is a restriction of graceful labelings for bipartite graphs. Theorem 5.12 has been recently generalized by López et al. in [30].

5.6 Concluding Remarks

An unpublished result of Erdős states that almost all graphs are not graceful (see [23]). However several families of graphs have been proven to admit a graceful labelings. Furthermore, Acharya has proven in [1] that every graph is an induced subgraph of a graceful graph. In this chapter we have revised several techniques that have been used in the study of graceful graphs. In particular we have concentrated on complete graphs and their relations with Golomb rulers, complete bipartite graphs, cycles, and mainly in trees, due to the importance of the graceful tree conjecture.

We have discussed techniques that allow to create new graceful trees from old ones using appropriate shifting of edges. In fact several other results have been obtained using this simple but powerful technique. In particular we want to pay special attention next to the following results and techniques that have not been discussed in the chapter, but that we feel that are interesting.

Mavronicolas and Michael described also in [31] very interesting techniques in order to construct graceful trees. However, it seems that so far no technique has been powerful enough in order to solve the graceful tree conjecture. Some authors have attacked the problem finding graceful labelings of trees one by one, with the aid of computers, however no counterexample for the conjecture has been found and, as a result of these efforts we have the following result (see for instance [3, 18, 26]).

Theorem 5.13 *Every tree of order up to 35 is graceful.*

Therefore, in spite of all efforts we may say that a final solution for the graceful tree conjecture seems to be far away so far. Although this conjecture is the most famous one in the world of graph labelings, there are also other conjectures that have been very much studied and that seem to be as difficult as the graceful tree conjecture. An example of such a conjecture is the Truszczyński's conjecture. Truszczyński studied in [42] the graceful properties of several classes of unicyclic graphs, and this study let him to conjecture the following.

Conjecture 5.3 (Truszczyński's Conjecture) All unicyclic graphs except C_n, where $n \equiv 1$ or $2 \pmod 4$, are graceful.

Again, a final solution to this conjecture seems to be very far. Many other questions and many other families of graphs have been considered when studying graceful labelings. For instance, Bagga et al. [9] have designed an algorithm for obtaining all graceful labelings of cycles. Aldred et al. [4] have considered the problem of counting different labelings of the path P_n, they have obtained the following result.

Theorem 5.14 *The number of graceful labelings of P_n is at least $(5/3)^n$.*

This bound has been improved by Adamaszek [2] to 2.7^n. Other families of graceful graphs have also been considered in the literature. For instance, Delorme et al. proved in [16] a conjecture of Bodenendieck [14].

Theorem 5.15 ([16]) *The graph consisting of a cycle with a chord is graceful.*

Delorme et al. also showed in [16] that other families of graphs such as n copies of the complete graph on four vertices having one edge in common is graceful. In the same paper they proved that n copies of the cycle C_4 having one edge in common is also graceful when $n + 1$ is not a multiple of 4. Many other families can be found in the literature. The interested reader can consult the following two surveys for further information on graceful graphs [6, 20].

Acknowledgements The proofs from [10] and Fig. 5.9 are introduced with permission from [10], Elsevier, ©2006. The proofs from [27] are introduced with permission from [27], Elsevier, ©2001. We gratefully acknowledge permission to use [37] by its author. We also gratefully acknowledge permission to use [39] by the publisher of Congr. Numer. The proof from [40] is introduced with permission from [40], Elsevier, ©1997.

References

1. Acharya, B.D.: Construction of certain infinite families of graceful graphs from a given graceful graph. Def. Sci. J. **32**(3), 231–236 (1982)
2. Adamaszek, M.: Efficient enumeration of graceful permutations. J. Combin. Math. Combin. Comput. **87**, 191–197 (2013)
3. Aldred, R.E.L., Mckay, B.D.: Graceful and harmonius labelings of trees. Bull. Inst. Combin. Appl. **23**, 69–72 (1998)
4. Aldred, R.E.L., Siran, J., Siran, M.: A note on the number of graceful labelings of paths. Discrete Math. **261**, 27–30 (2003)
5. Aravamudhan, R., Murugan, M.: Numbering of the vertices of $K_{a,1,b}$ (unpublished)
6. Arumugam, S., Bagga, J.: Graceful labeling algorithms and complexity - a survey. J. Indones. Math. Soc, 1–9 (2011). Special edn.
7. Babcock, W.C.: Intermodulation interference in radio Systems/Frequency of occurrence and control by channel selection. Bell Syst. Tech. J. **31**, 63–73 (1953)
8. Bača, M., Lin, Y., Muntaner-Batle, F.A.: Super edge-antimagic labelings of the path-like trees. Util. Math. **73**, 117–128 (2007)
9. Bagga, J., Heinz, A., Majumder, M.M.: An algorithm for computing all graceful labelings of cycles. Congr. Numer. **186**, 57–63 (2007)
10. Balbuena, C., García-Vázquez, P., Marcote, X., Valenzuela, J.C.: Trees having an even or quasi even degree sequence are graceful. Appl. Math. Lett. **20**, 370–375 (2007). http://dx.doi.org/10.1016/j.aml.2006.04.020
11. Barrientos, C.: Difference vertex labelings. Ph.D. thesis, Universitat Politècnica de Catalunya (2004)
12. Bermond, J.C.: Graceful graphs, radio antennae, and French windmills. In: Graph Theory and Combinatorics, pp. 18–37. Pitman Publishing Ltd., London (1979)
13. Bloom, G.S.: A counterexample to a theorem of S. Piccard. J. Combin. Theory (A) **22**, 378–379 (1977)
14. Bodendiek, R., Schumacher, H., Wegner, H.: Uber graziose graphen. Math. Phys. Semester-berichte **24**, 103–106 (1977)
15. Burzio, M., Ferrarese, G.: The subdivision graph of a graceful tree is a graceful tree. Discrete Math. **181**, 275–281 (1998)
16. Delorme, C., Maheo, M., Thuillier, H., Koh, K.M., Teo, H.K.: Cycles with a chord are graceful. J. Graph Theory **4**, 409–415 (1980)
17. El-Zanati, S.I., Kenig, M.J., Eynden, C.V.: Near α-labelings of bipartite graphs. Aust. J. Combin. **21**, 275–285 (2000)
18. Fang, W.: A computational approach to the graceful tree conjecture (2010). http://arxiv.org/abs/1003.3045
19. Gallian, J.A.: Living with the labeling disease for 25 years. J. Indones. Math. Soc. 54–88 (2011). Special edn.
20. Gallian, J.A.: A dynamic survey of graph labeling. Electron. J. Combin. **19**(DS6) (2016)
21. Gardner, M.: Mathematical games: the graceful graphs of Solomon Golomb, or how to number a graph parsimoniously. Sci. Am. **226**(3/4/6), 108–112; 104; 118 (1972)
22. Gnanajothi, R.B.: Topics in Graph Theory. Ph.D. thesis, Madurai Kamaraj University (1991)
23. Golomb, S.W.: How to number a graph. In: Graph Theory and Computing, pp. 23–37. Academic, New York (1972)
24. Grannell, M., Griggs, T., Holroy, F.: Modular gracious labelings of trees. Discrete Math. **231**, 199–219 (2001)
25. Hegde, S., Shetty, S.: On graceful trees. Appl. Math. E-Notes **2**, 192–197 (2002)
26. Horton, M.: Graceful trees: statistics and algorithms. Ph.D. thesis, University of Tasmania (2003)
27. Hrnčiar, P., Haviar, A.: All trees of diameter five are graceful. Discrete Math. **233**, 133–150 (2001). http://dx.doi.org/10.1016/S0012-365X(00)00233-8

28. IBM: http://www.research.ibm.com/people/s/shearer/grtab.html. Accessed June (2016)
29. Koh, K.M., Rogers, D.G., Tan, T.: On graceful trees. Nanta Math. **10**(2), 207–211 (1997)
30. López, S.C., Muntaner-Batle, F.A.: A new application of the $\otimes_h$-product to α-labelings. Discrete Math. **338**, 839–843 (2015)
31. Mavronicolas, M., Michael, L.: A substitution theorem for graceful trees and its applications. Discrete Math. **309**, 3757–3766 (2009)
32. Mixton, W.: cited in [21]
33. Muntaner-Batle, F.A., Rius-Font, M.: On the structure of path-like trees. Discuss. Math. Graph Theory **28**, 249–265 (2008)
34. Piccard, S.: Sur les ensembles de distances des ensembles de points d'un espace euclidien (1939)
35. Ringel, G.: Problem 25. In: Theory of Graphs and its Application (Proceedings of Symposium, Smolenice 1963), p. 162. Nakl. CSAV, Praha (1964)
36. Robinson, J.P., Bernstein, A.J.: A class of binary recurrent codes with limited error propagation. IEEE Trans. Inf. Theory **IT-13**, 106–113 (1967)
37. Rosa, A.: On certain valuations of the vertices of a graph. In: Gordon, N. Breach, D. Paris (eds.) Theory of Graphs (International Symposium, Rome, July 1966), pp. 349–355. Gordon and Breach/Dunod, New York/Paris (1967)
38. Sidon, S.: Ein statz über trigonometrishe polynome und seine anwendung in der thorie der Fourier Reihen. Math. Ann. **106**, 536–539 (1932)
39. Simmons, G.J.: Synch-sets: a variant of difference sets. Congr. Numer. **10**, 625–645 (1974)
40. Snevily, H.S.: New families of graphs that have α-labelings. Discrete Math. **170**, 185–194 (1997). http://dx.doi.org/10.1016/0012-365X(95)00159-T
41. Stanton, R., Zarnke, C.: Labeling of balanced trees. In: Proceedings of 4th Southeast Conference on Combinatorics Graph Theory and Computing, pp. 479–495. Utilitas Mathematica, Boca Raton (1973)
42. Truszczyński, M.: Graceful unicyclic graphs. Demonstatio Math. **17**, 377–387 (1984)

Chapter 6
The ⊗-Product of Digraphs: Second Type of Relations

6.1 Introduction

Since the beginning of graph labelings, researchers interested in this topic have dedicated their efforts mainly on finding techniques to prove the existence of particular families of graphs admitting some specific types of labeling. However, very few general techniques are known in order to create labelings of graphs.

This chapter is devoted to generate different kinds of labelings using a product of digraphs that was introduced by Figueroa-Centeno et al. [16]. This product is not only a generalization of the direct product of digraphs but also a generalization of the permutation voltage assignment. The key point of this new technique is to use super edge-magic labelings. From super edge-magic labelings we will see that we may obtain many other labelings, not only using the relations established in Sect. 3.1, but also using the product defined by Figueroa-Centeno et al. itself. This shows that super edge-magic labelings are a very powerful link among labelings, at least at two different levels. One level being the immediate relations already discussed, and the other level, being the relations existing between super edge-magic labelings and other labelings by means of this product.

Moreover, this product is used in order to find lower bounds for the number of nonisomorphic labelings of some particular graphs and to study the set of possible magic sums for some families of graphs, as for instance, cycles.

Since we may allow loops, graphs considered in this chapter are not necessarily simple. However, graphs with multiple edges will only be considered in very few occasions. We denote by L a loopgraph, that is a graph of order 1 and size 1. Let C_n be the cycle of order n. We denote by C_n^+ and by C_n^- the two possible strong orientations of the cycle C_n. We use the expression $\overrightarrow{G}$ to denote an oriented graph obtained from G and the expression und(D) denotes the underlying graph of a digraph D. For integers $m \le n$, let $[m, n] = \{m, m + 1, ..., n\}$.

In this chapter, when we say that a digraph is labeled with a particular labeling, we mean that its underlying graph has such a labeling.

© The Author(s) 2017
S.C. López, F.A. Muntaner-Batle, *Graceful, Harmonious and Magic Type Labelings*,
SpringerBriefs in Mathematics, DOI 10.1007/978-3-319-52657-7_6

6.2 The ⊗$_h$-Product of Digraphs

Definition 6.1 ([16]) Let D be a digraph and let Γ be a family of digraphs such that $V(F) = V$, for every $F \in \Gamma$. Consider any function $h : E(D) \longrightarrow \Gamma$. Then the product $D \otimes_h \Gamma$ is the digraph with vertex set the cartesian product $V(D) \times V$ and $((a, x), (b, y)) \in E(D \otimes_h \Gamma)$ if and only if $(a, b) \in E(D)$ and $(x, y) \in E(h((a, b)))$. When $|\Gamma| = 1$, we just write $D \otimes \Gamma$.

Consider the following two examples.

Example 6.1 Let D be the digraph with vertex set $\{1, 2\}$ and arc set $\{(1, 1), (1, 2)\}$. Let F_1 and F_2 be the digraphs on $V = [1, 3]$, such that $E(F_1) = \{(1, 2), (2, 3), (3, 1)\}$ and $E(F_2) = \{(1, 1), (2, 2), (3, 3)\}$. Let $h : E(D) \longrightarrow \{F_1, F_2\}$ be the function defined by $h((1, 1)) = F_1$ and $h((1, 2)) = F_2$. Then $D \otimes_h \Gamma$ is the digraph that appears in Fig. 6.1.

Example 6.2 Let D be the digraph with vertex set $\{1, 2\}$ and arc set $\{(1, 1), (1, 2), (2, 2)\}$. Let F_1 and F_2 be the digraphs on $V = [1, 5]$, such that $E(F_1) = \{(1, 4), (4, 2), (2, 5), (5, 3), (3, 1)\}$ and $E(F_2) = \{(1, 5), (5, 3), (3, 4), (4, 1), (2, 2)\}$. Let $h : E(D) \longrightarrow \{F_1, F_2\}$ be the function defined by $h((1, 1)) = h((2, 2)) = F_1$, and $h((1, 2)) = F_2$. Then $D \otimes_h \Gamma$ is the digraph that appears in Fig. 6.2, where due to space reasons, for each vertex the second coordinate appears as a subscript of the first one.

In what follows, we simplify the notation. Instead of $h((a, x))$ we write $h(a, x)$.

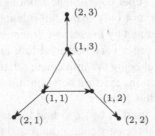

Fig. 6.1 An example of the ⊗$_h$-product

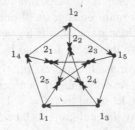

Fig. 6.2 An orientation of the Petersen graph obtained using the ⊗$_h$-product

All the graphs ω_n described in the proof of Theorem 3.7 can be obtained by means of the $\otimes_h$-product. See the next exercise.

Exercise 6.1 Show that every element ω_n, for $n \geq 1$, can be obtained by means of the $\otimes_h$-product of a path of length n with a loop attached at each vertex and the family of 1-regular digraphs of order 5. (For similar results see [30].)

6.2.1 The Adjacency Matrix

A different way to understand this product is by means of the adjacency matrices of the digraphs involved, and the adjacency matrix of the product itself. Let $A(D)$ and $A(F)$ be the adjacency matrices of D and $F \in \Gamma$, respectively, when the vertices of D are indexed as $V(D) = \{a_1, a_2, \ldots, a_m\}$ and the vertices of F as $V = \{x_1, x_2, \ldots, x_n\}$. Let the vertices of $D \otimes_h \Gamma$ be indexed as

$$\{(a_1, x_1), \ldots, (a_1, x_n), (a_2, x_1), \ldots, (a_m, x_n)\}.$$

Then, the adjacency matrix of $D \otimes_h \Gamma$, $A(D \otimes_h \Gamma)$, is obtained by multiplying every 0 entry of $A(D)$ by the $|V| \times |V|$ null matrix and every 1 entry of $A(D)$ by $A(h(a, b))$, where (a, b) is the arc related to the corresponding 1 entry. Notice that when h is constant, the adjacency matrix of $D \otimes_h \Gamma$ is just the classical Kronecker product $A(D) \otimes A(h(a, b))$.

Example 6.3 The adjacency matrix of the resulting digraph in Example 6.2, according to the vertex indexing

$$\{(1, 1), (1, 2), \ldots, (1, 5), (2, 1), \ldots, (2, 5)\},$$

appears below. It has been divided into four parts in order to better appreciate the adjacency matrices of all the elements involved in the product.

$$A(D \otimes_h \Gamma) = \left(\begin{array}{ccccc|ccccc} 0 & 0 & 0 & 1 & 0 & 0 & 0 & 0 & 0 & 1 \\ 0 & 0 & 0 & 0 & 1 & 0 & 1 & 0 & 0 & 0 \\ 1 & 0 & 0 & 0 & 0 & 0 & 0 & 0 & 1 & 0 \\ 0 & 1 & 0 & 0 & 0 & 1 & 0 & 0 & 0 & 0 \\ 0 & 0 & 1 & 0 & 0 & 0 & 0 & 1 & 0 & 0 \\ \hline 0 & 0 & 0 & 0 & 0 & 0 & 0 & 0 & 1 & 0 \\ 0 & 0 & 0 & 0 & 0 & 0 & 0 & 0 & 0 & 1 \\ 0 & 0 & 0 & 0 & 0 & 1 & 0 & 0 & 0 & 0 \\ 0 & 0 & 0 & 0 & 0 & 0 & 1 & 0 & 0 & 0 \\ 0 & 0 & 0 & 0 & 0 & 0 & 0 & 1 & 0 & 0 \end{array}\right)$$

6.2.2   Some Structural Properties

In this section, we study how different types of digraphs are obtained by the $\otimes_h$-product. Let Σ_n be the set of all 1-regular digraphs of order n. Without loss of restriction we can assume that the vertices of Σ_n are the elements in $[1, n]$. Consider a digraph $D \in \Sigma_n$ and let $j \in V(D)$, we denote by $j +_D 1$ the only vertex $k \in V(D)$ with $(j, k) \in E(D)$. Similarly, we denote by $j -_D 1$ the only vertex $i \in V(D)$ with $(i, j) \in E(D)$. Lemma 6.1 found next, will be useful in order to prove Theorem 6.1, which describes how the product behaves when multiplying an oriented tree by the set Σ_n.

Lemma 6.1 ([16])   Let T be a rooted tree with root $a \in V(T)$. For each function $h : E(\overrightarrow{T}) \rightarrow \Sigma_n$ there exist n labelings of $\overrightarrow{T}$, namely $l_i(h) : V(\overrightarrow{T}) \rightarrow [1, n]$, $i \in [1, n]$ such that:

  (i) $l_i(h)(a) = l_j(h)(a)$ if and only if $i = j$.
 (ii) If $xy \in E(T)$ and $d(a, x) < d(a, y)$ then

  – either $(x, y) \in E(\overrightarrow{T})$ and $(l_i(h)(x), l_i(h)(y)) \in E(h(x, y))$, or
  – $(y, x) \in E(\overrightarrow{T})$ and $(l_i(h)(y), l_i(h)(x)) \in E(h(y, x))$.

(iii) For all $x \in V(\overrightarrow{T})$, we obtain $\cup_i l_i(h)(x) = [1, n]$.
 (iv) If $h \neq h'$, then there exists i with $l_i(h) \neq l_j(h')$, for each $j \in [1, n]$.

Proof Fix a function $h : E(\overrightarrow{T}) \rightarrow \Sigma_n$. For each $i \in [1, n]$ the labeling $l_i(h)$ is defined recursively as follows: we label the root $l_i(h)(a) = i$. Assume that we have labeled the vertices of T up to level $k - 1$ and that z belongs to level k with $yz \in E(T)$ such that y belongs to level $k - 1$. If $(y, z) \in E(\overrightarrow{T})$ then $l_i(h)(z) = l_i(h)(y) +_{h(y,z)} 1$. If $(z, y) \in E(\overrightarrow{T})$, then $l_i(h)(z) = l_i(h)(y) -_{h(z,y)} 1$.

Clearly, each labeling $l_i(h)$ is defined by the label of vertex a. By construction, these labelings meet properties (i) and (ii). Also, since all elements in Σ_n are 1-regular digraphs, property (iii) holds for each labeling $l_i(h)$. If $h \neq h'$, then let $(r, s) \in E(h(x, y)) \setminus E(h'(x, y))$, for some $(x, y) \in E(\overrightarrow{T})$. Let i with $l_i(h)(x) = r$. Then $s = l_i(h)(y)$ and $s \neq l_i(h')(y)$. Thus, $l_i(h) \neq l_i(h')$. For each $j \neq i$, we have $l_j(h')(a) = j$ and $l_i(h)(a) = i$. Hence, property (iv) and the lemma follow.   □

Next, we illustrate the labelings introduced in Lemma 6.1 in the following example.

Example 6.4 Let $\overrightarrow{T}$ be the tree defined by $V(\overrightarrow{T}) = \{a_1, a_2, a_3, a_4, a_5\}$ and $E(\overrightarrow{T}) = \{(a_1, a_2), (a_3, a_2), (a_2, a_4), (a_3, a_5)\}$. Let F_i be the directed cycles on $\{1, 2, 3, 4\}$, for $i = 1, 2, 3$ defined by $E(F_1) = \{(1, 2), (2, 3), (3, 4), (4, 1)\}$, $E(F_2) = \{(1, 4), (4, 3), (3, 2), (2, 1)\}$ and $F_3 = \{(1, 2), (2, 4), (4, 3), (3, 1)\}$. Consider the function $h : E(\overrightarrow{T}) \rightarrow \Sigma_4$ defined by $h(a_1, a_2) = F_1$, $h(a_2, a_4) = F_2$, $h(a_3, a_2) = F_3$ and $h(a_3, a_5) = F_1$. Then, the labelings $l_i(h)$ of $\overrightarrow{T}$, with root in a_2, are shown in Fig. 6.3.

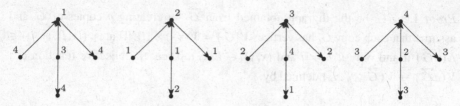

Fig. 6.3 The labelings $l_i(h)$, for $i \in [1, 4]$, introduced on Lemma 6.1

In the next theorem, we use Lemma 6.1 in order to establish an isomorphism between the digraphs $n\overrightarrow{T}$ and $\overrightarrow{T} \otimes_h \Sigma_n$, where h is any function with domain $E(\overrightarrow{T})$ and range in Σ_n.

Theorem 6.1 ([16]) *Let T be a tree. Consider any function $h : E(\overrightarrow{T}) \to \Sigma_n$. Then, $\overrightarrow{T} \otimes_h \Sigma_n \cong n\overrightarrow{T}$.*

Proof Let $n\overrightarrow{T}$ be the digraph obtained from $\overrightarrow{T}$ by creating n disjoint copies of $\overrightarrow{T}$ and assume that each copy $\overrightarrow{T_i}$ has vertex set $V(\overrightarrow{T_i}) = V(\overrightarrow{T}) \times \{i\}$. That is, $((x, i), (y, j)) \in E(n\overrightarrow{T})$ if and only if $i = j$ and $(x, y) \in E(\overrightarrow{T})$. Fix any function $h : E(\overrightarrow{T}) \to \Sigma_n$. By definition, the vertices of $\overrightarrow{T} \otimes_h \Sigma_n$ are $\{(x, j) : x \in V(\overrightarrow{T}), j \in [1, n]\}$. Thus, $V(n\overrightarrow{T}) = V(\overrightarrow{T} \otimes_h \Sigma_n)$. Let us prove now that the correspondence $(x, i) \leftrightarrow (x, l_i(x))$ is an isomorphism between the digraphs $n\overrightarrow{T}$ and $\overrightarrow{T} \otimes_h \Sigma_n$. If $((x, i), (y, j))$ is an arc in $n\overrightarrow{T}$, then $i = j$ and $(x, y) \in E(\overrightarrow{T})$. Thus, by definition of the labelings $l_i(h)$, we obtain $(l_i(h)(x), l_i(h)(y)) \in E(h(x, y))$, and, by definition of the $\otimes_h$-product, $((x, l_i(h)(x)), (y, l_i(h)(y))) \in E(\overrightarrow{T} \otimes_h \Sigma_n)$. Similarly, if $((x, l_i(h)(x)), (y, l_j(h)(y))) \in E(\overrightarrow{T} \otimes_h \Sigma_n)$, then $(x, y) \in E(\overrightarrow{T})$ and $(l_i(h)(x), l_j(h)(y)) \in E(h(x, y))$. Thus, by definition of the labelings $l_i(h)$, we obtain $i = j$. Therefore, $((x, i), (y, j)) \in E(n\overrightarrow{T})$. □

It is an easy exercise to generalize Lemma 6.1 to any oriented forest. Using this generalization, the next theorem can be easily obtained.

Theorem 6.2 ([16]) *Let F be an acyclic graph. Consider any function $h : E(\overrightarrow{F}) \to \Sigma_n$. Then, $\overrightarrow{F} \otimes_h \Sigma_n \cong n\overrightarrow{F}$.*

The union of, nonnecessarily acyclic, bipartite graphs can also be obtained using the $\otimes_h$-product. However, in this case, not all orientations of the edges will produce the same underlying graph, when applying the product.

Theorem 6.3 ([16]) *Let G be bipartite graph with stable sets V_1 and V_2. Let $\overrightarrow{G}$ be an orientation of G such that if $(u, v) \in E(\overrightarrow{G})$ then $u \in V_1$. Consider any constant function $h : E(\overrightarrow{G}) \to \Sigma_n$. Then, $\overrightarrow{G} \otimes_h \Sigma_n \cong n\overrightarrow{G}$.*

Proof Let $n\vec{G}$ be the digraph obtained from $\vec{G}$ by creating n copies of $\vec{G}$ and assume that each copy $\vec{G_i}$ has vertices $V(\vec{G_i}) = V(\vec{G}) \times \{i\}$. That is, $((x, i), (y, j)) \in E(n\vec{G})$ if and only if $i = j$ and $(x, y) \in E(\vec{G})$. Then, the bijective function φ : $V(n\vec{G}) \to V(\vec{G} \otimes_h \Sigma_n)$ defined by:

$$\varphi(x, j) = \begin{cases} (x, j), & \text{if } x \in V_1, \\ (x, j +_F 1), & \text{if } x \in V_2, \end{cases}$$

where $F = h(u, v)$ and $(u, v) \in E(\vec{G})$, defines an isomorphism among the digraphs $n\vec{G}$ and $\vec{G} \otimes_h \Sigma_n$. $\square$

A known result in the area (see, for instance, [23]) is that the direct product of two strongly oriented cycles produces copies of a strongly oriented cycle, namely,

$$C_m^+ \otimes C_n^+ \cong \gcd(m, n)C_{\text{lcm}(m,n)}^+. \tag{6.1}$$

An extension of (6.1) was obtained by Ahmad et al. in [1].

Theorem 6.4 ([1]) *Let $m, n \in \mathbb{N}$ and consider the product $C_m^+ \otimes_h \{C_n^+, C_n^-\}$ where $h : E(C_m^+) \longrightarrow \{C_n^+, C_n^-\}$. Let g be a generator of a cyclic subgroup of $\mathbb{Z}_n$, namely $\langle g \rangle$, such that $|\langle g \rangle| = k$. Also let $r < m$ be a positive integer that satisfies the following congruence relation*

$$m - 2r \equiv g \pmod{n}.$$

If the function h assigns C_n^- to exactly r arcs of C_m^+, then the product

$$C_m^+ \otimes_h \{C_n^+, C_n^-\}$$

consists of exactly n/k disjoint copies of a strongly oriented cycle C_{mk}^+. In particular if $\gcd(g, n) = 1$, then $\langle g \rangle = \mathbb{Z}_n$ and if the function h assigns C_n^- to exactly r arcs of C_m^+ then

$$C_m^+ \otimes_h \{C_n^+, C_n^-\} \cong C_{mn}^+.$$

Proof Let $V(C_l^+) = \mathbb{Z}_l$ and $E(C_l^+) = \{(i, j) : j - i \equiv 1 \pmod{l}\}$, for $l = m, n$. Similarly, let $V(C_n^-) = \mathbb{Z}_n$ and $E(C_n^-) = \{(i, j) : i - j \equiv 1 \pmod{n}\}$. Consider any function h that assigns C_n^- to exactly r arcs of C_m^+. Also consider a directed subpath of $C_m^+ \otimes_h \{C_n^+, C_n^-\}$ (possibly closed) that starts at $(0, 0)$:

$$((0, 0), (1, b_1)), ((1, b_1), (2, b_2)), \ldots$$

Then, the vertex that is in position mk in the path is a vertex of the form $(0, k(m-2r))$ (mod n). Recall that k is the smallest number that satisfies the following congruence relation $kg \equiv 0 \pmod{n}$ since $|\langle g \rangle| = k$. Therefore we cannot close an oriented cycle of smaller length than mk. A similar reasoning will give us the remaining cycles of length mk. □

It is interesting to notice that the previous result was generalized recently in [29].

6.2.3 The $\otimes_h$-Product as a Generalization of Voltage Assignments

Voltage graphs were introduced in the context of topological graph theory, which studies embeddings of graphs on different surfaces, such as the sphere, the torus, etc. It is interesting enough to notice that there is a close relationship between the $\otimes_h$-product, that will be used in order to construct labelings of graphs, and voltage graphs. Therefore, from this point of view, we have a connection between topological graph theory and graph labelings. In fact, it turns out that the $\otimes_h$-product of digraphs can be though as a generalization of the permutation voltage-graph introduced by Gross and Tucker in [21]. An excellent source of information on voltage graphs in general that can be consulted by the interested reader is [22].

Next, we introduce the concepts of permutation voltage assignment, base graph and permutation voltage graph.

Definition 6.2 Let G be a graph whose edges have all been assigned plus and minus directions. A *permutation voltage assignment* for G is a function α from the plus-directed edges of G into a symmetric group $\mathfrak{S}_n$. The permutation voltage on a minus-directed edge e^- is understood to be the inverse permutation of the voltage on e^+. The graph G is called the *base graph* and the subscripted pair $\langle G, \alpha \rangle_n$ is called a *permutation voltage graph*.

To each such a permutation voltage graph there is associated a *permutation derived graph* G^α, whose vertex set is the cartesian product $V(G) \times [1, n]$ and whose edge set is the cartesian product $E(G) \times [1, n]$. If $e = uv \in E(G)$, $e^+ = (u, v)$ and π is the permutation assigned to e^+, then $(u, i)(v, \pi(i))$ belongs to $E(G^\alpha)$, for every $i \in [1, n]$.

The relation between permutation derived graphs and the graphs obtained when using the $\otimes_h$-product is established in the next result.

Lemma 6.2 *Let D be the oriented digraph obtained from G when considering all plus-directed edges. For each permutation $\pi \in \mathfrak{S}_n$, consider the digraph F_π on $V = [1, n]$ whose arc set is $\{(i, \pi(i)) : i \in [1, n]\}$. Then,*

$$und(D \otimes_h \{F_\pi\}_{\pi \in \mathfrak{S}_n}) \cong G^\alpha,$$

where h is the function defined by $h(e^+) = F_{\alpha(e^+)}$.

Remark 6.1 Example 6.1 and Fig. 6.1 can also be found in [22, Chap. 2], as an illustration of the voltage-graph construction.

6.3 SEM Labelings

The first paper that uses the $\otimes_h$-product for constructing labelings is [16]. Almost all results contained in [16] use as the second factor of the product the set of SEM 1-regular digraphs of odd order n, that is denoted by $\mathscr{S}_n$. It turns out that many of the results in [16] also hold when instead of considering the set of SEM 1-regular digraphs, we consider families of SEM labeled digraphs with size equal to order, provided that the magic sum for each element of the family is constant.

A super edge-magic labeled digraph F is in the set $\mathscr{S}_n^k$ if $|V(F)| = |E(F)| = n$ and the minimum sum of the labels of the adjacent vertices is equal to k.

Notice that, by Corollary 3.4, the magic sum of a super edge-magic 1-regular digraph of order n is $(5n + 3)/2$. Thus, the minimum sum of the labels of adjacent vertices is equal to $(n + 3)/2$. Hence, we have $\mathscr{S}_n \subset \mathscr{S}_n^{(n+3)/2}$ and, therefore, the family $\mathscr{S}_n^k$ is a generalization of the family $\mathscr{S}_n$.

Theorem 6.5 ([31]) *Assume that D is any (super) edge-magic digraph and h is any function $h : E(D) \rightarrow \mathscr{S}_n^k$. Then $und(D \otimes_h \mathscr{S}_n^k)$ is (super) edge-magic.*

Proof We rename the vertices of D after the labels of a (super) edge-magic labeling f. Then, we define the labels of the product as follows:

1. If $(i,j) \in V(D \otimes_h \mathscr{S}_n^k)$, we assign to the vertex the label: $n(i - 1) + j$.
2. If $((i,j),(i',j')) \in E(D \otimes_h \mathscr{S}_n^k)$, we assign to the arc the label: $n(e - 1) + (k + n) - (j + j')$, where e is the label of (i, i') in D.

Notice that, since each element F of $\mathscr{S}_n^k$ is labeled with a super edge-magic labeling with minimum sum of the adjacent vertices equal to k, we have

$$\{(k + n) - (j + j') : (j,j') \in E(F)\} = [1, n].$$

Thus, the set of labels in $D \otimes_h \mathscr{S}_n^k$ covers all elements in $[1, n(|V(D)| + |E(D)|)]$. Moreover, for each arc $((i,j)(i',j')) \in E(D \otimes_h \mathscr{S}_n^k)$, coming from an arc $e = (i, i') \in E(D)$ and an arc $(j,j') \in E(h(i, i'))$, the sum of the labels is constant and is equal to:

$$n(i + i' + e - 3) + k + n. \tag{6.2}$$

That is, $n(\sigma_f - 3) + k + n$, where σ_f denotes the magic sum of the labeling f. We also notice that if the digraph D is super edge-magic then the vertices of $D \otimes_h \mathscr{S}_n^k$ receive the smallest labels.   □

Example 6.5 The orientation of the Petersen graph obtained in Example 6.2 (see also Fig. 6.2) is the product of a SEM labeled digraph with a family of SEM labeled digraphs. Thus, by replacing each vertex (i,j) with $5(i - 1) + j$ we obtain a SEM labeling of it. See Fig. 6.4.

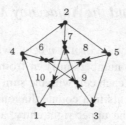

Fig. 6.4 A SEM labeling of an orientation of the Petersen graph obtained from the $\otimes_h$-product of SEM digraphs

Notice that, if D is bipartite then, the digraph obtained from the product, $D \otimes_h \mathscr{S}_n^k$ is also bipartite. Moreover, if D admits a special super edge-magic labeling, then the labeling shown in the proof of Theorem 6.5 is a special super edge-magic labeling of $D \otimes_h \mathscr{S}_n^k$. Thus, we obtain the next result.

Corollary 6.1 *Let D be a special super edge-magic digraph and let $h : E(D) \to \mathscr{S}_n^k$ be any function. Then, $(D \otimes_h \mathscr{S}_n^k)$ is special super edge-magic.*

From the proof of Theorem 6.5, when $k = (n+3)/2$, we also obtain the following results.

Proposition 6.1 ([32]) *Let $\check{f}$ be the edge-magic labeling of the graph und$(D \otimes_h \mathscr{S}_n)$ obtained in Theorem 6.5 from a labeling f of D. Then the magic sum of $\check{f}$, $\sigma_{\check{f}}$, is given by the formula*

$$\sigma_{\check{f}} = n(\sigma_f - 3) + \frac{3n + 3}{2}, \tag{6.3}$$

where σ_f is the magic sum of f.

Corollary 6.2 ([32]) *Let D be an edge-magic digraph and assume that there exist two edge-magic labelings of D, f, and g, such that $\sigma_f \neq \sigma_g$. If we denote by $\check{f}$ and $\check{g}$ the edge-magic labelings of the graph und$(D \otimes_h \mathscr{S}_n)$ when using the edge-magic labelings f and g of D respectively, then we get*

$$|\sigma_{\check{f}} - \sigma_{\check{g}}| \geq 3.$$

Proof Since $\sigma_f \neq \sigma_g$, we get the inequality $|\sigma_f - \sigma_g| \geq 1$. Thus, by using (6.3), we obtain $|\sigma_{\check{f}} - \sigma_{\check{g}}| = |n(\sigma_f - \sigma_g)| \geq 3$. □

6.3.1 SEM Labelings and the Adjacency Matrix

Suppose that we construct a square of $n \times n$ cells, with the rows and the columns labeled consecutively with the numbers in $[1, n]$ from left to right and from top to bottom, respectively. If we fill each cell with the sum of the corresponding column and row labels, then all elements in a counterdiagonal, that is a diagonal running from the lower left entry to the upper right entry, are the same. Moreover, consecutive counterdiagonals have consecutive sums. From this remark, the following lemma comes easily.

Lemma 6.3 ([16]) *A digraph D is SEM if and only if the adjacency matrix (a_{ij}) of D obtained by relabeling the vertices of D after the corresponding labels of a SEM labeling has the following two properties:*

(a) *If $a_{ij} = a_{i'j'} = 1$, then $i + j \neq i' + j'$, whenever $(i, j) \neq (i', j')$ and*
(b) *the counterdiagonals $i + j = k$ of the adjacency matrix with some entry different from 0 are consecutive.*

Example 6.6 Figure 6.5 shows a table of 5×5 cells, in which all rows and columns are supposed to be labeled consecutively $1, 2, \ldots, 5$ from left to right and from top to bottom, respectively. Then, each cell is filled with the sum of the corresponding column and row. Besides this table, we add the adjacency matrix of the super edge-magic labeled oriented cycle C_5^+ that appears in Fig. 2.5b.

Remark 6.2 Let $|V| = n$. Let $A(D)$ and $A(F)$ be the adjacency matrices of D and $F \in \Gamma$, when the vertices of D and F are identified with the labels of a super edge-magic labeling. If we relabel each vertex of the product (i, j) with $n(i - 1) + j$, then the adjacency matrix of $D \otimes_h \Gamma$ is obtained by multiplying every 0 entry of $A(D)$ by the $|V| \times |V|$ null matrix and every 1 entry of $A(D)$ by $A(h(a, b))$, where (a, b) is the arc related to the corresponding 1 entry. Example 6.3 shows the adjacency matrix of $D \otimes_h \Gamma$. Thus, from this construction of the adjacency matrix of the product, we can obtain a super edge-magic labeling of the resulting digraph.

2	3	4	5	6
3	4	5	6	7
4	5	6	7	8
5	6	7	8	9
6	7	8	9	10

$$\begin{pmatrix} 0 & 0 & 0 & 1 & 0 \\ 0 & 0 & 0 & 0 & 1 \\ 1 & 0 & 0 & 0 & 0 \\ 0 & 1 & 0 & 0 & 0 \\ 0 & 0 & 1 & 0 & 0 \end{pmatrix}$$

Fig. 6.5 A table of order 5×5 and the adjacency matrix of a SEM labeled digraph of order 5

This interpretation of the product was, in fact, the original idea for obtaining (super) edge-magic labelings from the product of (super) edge-magic digraphs. The title of the paper were the $\otimes_h$ was introduced, *Labeling Generating Matrices* [16], points out this fact.

6.3.2 SEM Labelings of 2-Regular Graphs

As we have seen in Lemma 3.5, if a (p, q)-graph G is super edge-magic, then $q \leq 2p-3$. Moreover, by Corollary 3.5, if G is bipartite, then this bound can be improved to $q \leq 2p - 5$. These bounds imply that if an r-regular graph is super edge-magic, then $r \in [0, 3]$. It is clear that 0-regular graphs are of no interest in this context. Furthermore, 1-regular edge-magic graphs have been completely characterized in [28] by Kotzig and Rosa, under the name M-valuation. By interchanging the labels they provided, we obtain the next result.

Theorem 6.6 ([28]) *Let G be a 1-regular $(2n, n)$-graph. Then G is (super) edge-magic if and only n is odd.*

Proof By Corollary 3.4, if a 1-regular $(2n, n)$-graph is super edge-magic, then n is odd. Similarly, if a 1-regular $(2n, n)$-graph is edge-magic, then the sum of all induced edge sums should be divisible by n. Thus, since each element in $[1, 3n]$ appears once, n should divide $3n(3n + 1)/2$. That is, n should be odd. Let $V(G) = \{a_i, b_i\}_{i=1}^n$, $E(G) = \{a_i b_i\}_{i=1}^n$ and $n = 2k + 1$. Define $f : V(G) \cup E(G) \to [1, 3n]$ by $f(a_i) = i$,

$$f(b_i) = \begin{cases} 4k + 4 - 2i, & \text{if } 1 \leq i \leq k + 1, \\ 6k + 5 - 2i, & \text{if } k + 2 \leq i \leq 2k + 1, \end{cases}$$

$$f(a_i b_i) = \begin{cases} 5k + 2 + i, & \text{if } 1 \leq i \leq k + 1, \\ 3k + 1 + i, & \text{if } k + 2 \leq i \leq 2k + 1. \end{cases}$$

Then, $f(V(G)) = [1, 4k+2], f(E(G)) = [4k+3, 6k+3]$ and $f(a_i)+f(b_i)+f(a_i b_i) = 9k + 6$. $\qquad\square$

Therefore, it seems to be a natural question to ask which 2-regular graphs are super edge-magic. Moreover, super edge-magic labelings of 2-regular graphs are a key point in order to develop further results, not only using the $\otimes_h$-product, but also for direct connections with other labelings. The next result is an example. Recall that a vertex-magic total labeling (VMTL) of a graph G is a bijective function $f : V(G) \cup E(G) \to [1, |V(G)| + |E(G)|]$ such that $f(x) + \sum_{y \in N_G(x)} f(xy) = k$, for some $k \in \mathbb{Z}$ and for every $x \in V(G)$. The VMTL is called strong (super) if the vertices are labeled with the largest (smallest) available integers.

Theorem 6.7 ([19]) *If G is a graph with a spanning subgraph H which possesses a strong vertex-magic total labeling and G − E(H) is even regular, then G also possesses a strong vertex-magic total labeling.*

Let G be a (p, q)-graph. By introducing the complement labeling of a given labeling f of G, namely $\bar{f}(v) = p + q + 1 - f(v)$, it is clear that any regular graph has a strong VMTL if and only if it has a super VMTL. Furthermore, a 2-regular graph G admits a vertex-magic total labeling if and only if G admits an edge-magic labeling of G, since in a 2-regular graph G the two notions coincide. Thus, the existence of super edge-magic labelings of 2-regular graphs has been studied in different papers [16, 19, 20, 24]. Figueroa-Centeno et al. in [17] conjectured that the graph $C_m \cup C_n \neq C_3 \cup C_4$ is super edge-magic if and only if $m + n$ is odd and greater than 1. Holden et al. went further into this conjecture when they conjectured in [24] that all 2-regular graphs of odd order are strong vertex total magic, excluding $C_3 \cup C_4$, $3C_3 \cup C_4$ and $2C_3 \cup C_5$.

Next, we cite some of the results in this context.

Theorem 6.8 ([17, 20]) *The 2-regular graph $C_3 \cup C_n$ is super edge-magic if and only if $n \geq 6$ and n is even.*

Theorem 6.9 ([17, 20, 24]) *The 2-regular graph $C_4 \cup C_n$ is super edge-magic if and only if $n \geq 5$ and n is odd.*

Theorem 6.10 ([17, 24]) *The 2-regular graph $C_5 \cup C_n$ is super edge-magic if and only if $n \geq 4$ and n is even.*

Theorem 6.11 ([17]) *If m is even with $m \geq 6$ and n is odd with $n \geq m/2 + 2$, then the 2-regular graph $C_m \cup C_n$ is super edge-magic.*

By using the $\otimes_h$-product, we present an extension of the previous results to the family $C_m \cup C_n$ with m even, n odd, and m being a multiple of n. The first result proves that the union of a loop with a cycle of even order is super edge-magic. Then, using the $\otimes_h$-product, we enlarge the class of known 2-regular super edge-magic graphs with exactly two components. Furthermore, by Lemma 3.1 in Sect. 3.1, we also enlarge the class of known harmonious 2-regular graphs with exactly two components.

By parity reasons, the union of a loop with a cycle of odd length is not super edge-magic. The next lemma shows a super edge-magic labeling for the union of a loop with a cycle of even length.

Lemma 6.4 ([31]) *Let m be an even integer, with $m \geq 4$. Then $C_m \cup L$ is super edge-magic.*

Proof Let $V(C_m \cup L) = \{v_i\}_{i=0}^{m}$ and $E(C_m \cup L) = \{v_i v_{i+1}\}_{i=1}^{m-1} \cup \{v_m v_1\} \cup \{v_0 v_0\}$. We distinguish two cases.

Case $m \equiv 0$ *(mod 4)*. We consider the labeling:

$$
f(v_i) = \begin{cases}
(i+1)/2, & i \text{ odd}, \\
i/2 + m/2, & i \text{ even and } i \leq m/2, \\
i/2 + m/2 + 1, & i \text{ even and } i > m/2, \\
m/2 + m/4 + 1, & i = 0.
\end{cases}
$$

Case $m \equiv 2$ *(mod 4)*. We consider the labeling:

$$
f(v_i) = \begin{cases}
(i+1)/2, & i \text{ odd and } i \leq m/2, \\
(i+1)/2 + m/2, & i \text{ odd and } i > m/2, \\
m + 1, & i = 2, \\
i/2 + m/2, & i \text{ even and } 2 < i \leq (m+2)/2, \\
i/2 + 1, & i \text{ even and } i > (m+2)/2, \\
(m+2)/4 + 1, & i = 0.
\end{cases}
$$

Notice that, in both cases, the labeling f assigns the labels from 1 up to $m + 1$ to the vertices and the induced edge sums are consecutive. Thus, by Lemma 2.1 the labeling f is super edge-magic. □

Two examples for the labelings provided in the proof of Lemma 6.4 are shown in Fig. 6.6.

Theorem 6.12 ([31]) *Let m be an even integer. If n is an odd divisor of m, with $m/n, n \geq 3$ and either $m > n(n-1)$ or $\gcd(n, m/n) = 1$, then the 2-regular graph $C_m \cup C_n$ is super edge-magic.*

Proof Let m and n be two positive integers, such that m is even, n is odd and $m = kn$, for some integer k. Since k is even, by Lemma 6.4, the graph $C_k \cup L$ is super edge-magic. Thus, by Theorem 6.5, for any function $h : E(C_k^+ \cup \overrightarrow{L}) \to \{C_n^+, C_n^-\}$, the graph $\mathrm{und}(C_k^+ \cup \overrightarrow{L}) \otimes_h \{C_n^+, C_n^-\}$ is super edge-magic, where the vertices of C_n are identified with the labels of a super edge-magic labeling. Let us see now that

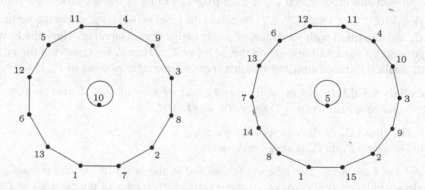

Fig. 6.6 $C_{12} \cup L$ and $C_{14} \cup L$ with the labeling provided in the proof of Lemma 6.4

there exists a function $h : E(C_k^+ \cup \overrightarrow{L}) \to \{C_n^+, C_n^-\}$ such that $\text{und}((C_k^+ \cup \overrightarrow{L}) \otimes_h \{C_n^+, C_n^-\}) \cong C_{kn} \cup C_n$. Notice that, by definition of the $\otimes_h$-product, we have

$$(C_k^+ \cup \overrightarrow{L}) \otimes_h \{C_n^+, C_n^-\} \cong (C_k^+ \otimes_{h_{|E(C_k^+)}} \{C_n^+, C_n^-\}) \cup (\overrightarrow{L} \otimes_{h_{|E(\overrightarrow{L})}} \{C_n^+, C_n^-\})$$

and that $\overrightarrow{L} \otimes_{h_{|E(\overrightarrow{L})}} \{C_n^+, C_n^-\} \cong h(E(\overrightarrow{L}))$. Hence, we only need to find a function $h_1 : E(C_k^+) \to \{C_n^+, C_n^-\}$ such that $\text{und}(C_k^+ \otimes_{h_1} \{C_n^+, C_n^-\}) \cong C_{kn}$.

Assume that $\gcd(k, n) > 1$, otherwise, considering a constant function, the result holds by (6.1). Since n is odd, we have $\langle 1 \rangle = \mathbb{Z}_n$ and the congruence relation

$$k - 2r \equiv 1 \ (mod \ n)$$

can be solved with $0 < r < k$. Therefore, by considering any function h_1 that assigns C_n^- to exactly r arcs of C_k^+, Theorem 6.4 implies that $\text{und}(C_k^+ \otimes_{h_1} \{C_n^+, C_n^-\}) \cong C_{kn}$. □

In [35], the following question, which is still open, was proposed.

Question 6.1 Characterize the set of 2-regular harmonious graphs.

Using Lemma 3.1 on Sect. 3.1 together with Theorem 6.12, we can extend the partial answer given in [17] to provide more inside for this question.

Corollary 6.3 ([31]) *Let m be an even integer. If n is an odd divisor of m, with $m/n, n \geq 3$ and either $m > n(n-1)$ or $\gcd(n, m/n) = 1$, then the 2-regular graph $C_m \cup C_n$ is harmonious.*

By Theorem 6.12 and Lemma 6.4, we also obtain the following:

Corollary 6.4 ([31]) *Let m be an even integer. If n is an odd divisor of m, with $m/n, n \geq 3$ and either $m > n(n-1)$ or $\gcd(n, m/n) = 1$, then the graph $P_m \cup C_n$ is super edge-magic.*

Proof Let $k = m/n$. The labeling introduced in the proof of Lemma 6.4 assigns the lower induced sum between adjacent vertices to the cycle C_k. By definition of the $\otimes_h$-product and the induced super edge-magic labeling of the $\otimes_h$-product, a vertex $(i, j) \in V((C_k^+ \cup \overrightarrow{L}) \otimes_h \{C_n^+, C_n^-\})$ receives the label $n(i-1) + j$, where the vertices of C_n are identified with the labels of a super edge-magic labeling. Thus, the lower edge induced sum will appear on the edges of C_m. Hence, by removing the edge with smallest induced sum, we obtain a super edge-magic labeling of $P_m \cup C_n$. □

Corollary 6.5 ([31]) *Let $m \equiv 0 \ (mod \ 4)$, and let n be an odd divisor of m, with $n \geq 3$ and either $m > n(n-1)$ or $\gcd(n, m/n) = 1$. Then,*

(i) The graph $C_m \cup P_n$ is super edge-magic.
(ii) The graph $P_m \cup P_n$ is super edge-magic.

Proof Let $k = m/n$. The labeling introduced in the proof of Lemma 6.4 assigns the lower induced sum between adjacent vertices to an edge of the cycle C_k and the bigger one to the loop. Thus, by definition of the $\otimes_h$-product and the induced super edge-magic labeling of the $\otimes_h$-product, the lower and the bigger edge induced sums appear among the edges of C_m and the edges of C_n, respectively. By removing the edge with biggest induced sum, we obtain a super edge-magic labeling of $C_m \cup P_n$, which proves (i). Finally, by removing the edge with biggest induced sum together with the edge with lowest induced sum from $C_m \cup C_n$, we obtain (ii).   □

6.4 Nonisomorphic Labelings

Let G be a (S)(S)EM labeled graph, for which the vertices take the name of their corresponding labels, and let $h : E(\overrightarrow{G}) \to \mathscr{S}_n$ be any function. We denote by $l(h)$ the labeling of und($\overrightarrow{G} \otimes_h \mathscr{S}_n$) induced by the labeling of D and of $\{h(x, y) : (x, y) \in E(\overrightarrow{G})\}$, namely, $l(h)(x, j) = n(x - 1) + j$.

Lemma 6.5 ([16])   *Let T be any (S)(S)EM labeled tree for which each vertex takes the name of its label and let $h, h' : E(\overrightarrow{T}) \to \mathscr{S}_n$. Then, $l(h)$ and $l(h')$ are isomorphic labelings of nT if and only if $l(h) = l(h')$. That is, $l(h)$ and $l(h')$ are isomorphic labelings of nT if and only if $h = h'$.*

Proof We view the tree as a rooted tree, and we let the root to be the vertex labeled 1. According to that, the vertices of $\overrightarrow{T} \otimes_h \mathscr{S}_n$ can be described as $\{(x, l_i(h)(x)) : x \in V(T), i \in [1, n]\}$, where $l_i(h)$ are the labelings introduced in Lemma 6.1. We want to show that the automorphism g defined by the rule, $g(x, l_i(h)(x)) = (y, l_j(h)(y))$ if and only if $l(h)(x, l_i(h)(x)) = l(h')(y, l_j(h)(y))$ is the identity function. By definition, the two labelings coincide on the roots of the trees. Assume that the labels coincide on the vertices which are at distance at most $k - 1$ from the root of its component of the forest. Let $(x, l_i(h)(x))$ be a vertex at distance k from the root of its component, and let $(y, l_j(h)(y))$ be a vertex with $g(x, l_i(h)(x)) = g(y, l_j(h)(y))$. Since g is an automorphism, it follows that $(x, l_i(h)(x))$ and $(y, l_j(h)(y))$ belong to the same component. Thus, $i = j$ and they are at the same distance from the root. On the other hand, since $l(h)(x, l_i(h)(x)) \in (n(x - 1), nx]$ and $l(h')(y, l_j(h)(y)) \in (n(y - 1), ny]$, we have $x = y$. In particular, equality $l_i(h) = l_i(h')$ holds for each $i \in [1, n]$ and by Lemma 6.1 (iv) we obtain $h = h'$.   □

We leave as an easy exercise the generalization of the previous result that is stated in the next corollary.

Corollary 6.6 ([16])   *Let F be any (S)(S)EM labeled forest and let $h, h' : E(\overrightarrow{F}) \to \mathscr{S}_n$. Then, $l(h)$ and $l(h')$ are isomorphic labelings of nF if and only if $l(h) = l(h')$. That is, $l(h)$ and $l(h')$ are isomorphic labelings of nF if and only if $h = h'$.*

Thus, we are able to obtain a lower bound for the number of nonisomorphic (S)(S)EM labelings of n copies of a (S)(S)EM forest, when n is odd.

Corollary 6.7 ([16])   *Let F be an (S)(S)EM acyclic graph of order m with p components, and let n be an odd integer. Then, the graph nF admits at least $|\mathscr{S}_n|^{m-p}$ nonisomorphic (S)(S)EM labelings.*

Proof By Corollary 6.6, there exist as many (S)(S)EM labelings as functions h : $E(\overrightarrow{F}) \to \mathscr{S}_n$. Thus, the result follows.   □

An improvement of the previous lower bound can be obtained when considering odd disjoint unions of complete graphs on two vertices. Let n be an odd integer. We denote by $N(n)$ the number of nonisomorphic (S)(S)EM labelings of the graph nK_2. By Corollary 6.7, we obtain $N(n) \geq |\mathscr{S}_n|$.

Theorem 6.13 ([16])   *Let m and n be odd integers. Then, the graph $(mn)K_2$ admits at least $\max\{N(n)|\mathscr{S}_m|^n, N(m)|\mathscr{S}_n|^m\}$ nonisomorphic (S)(S)EM labelings.*

Proof By Corollary 6.7, for every (S)(S)EM labelings of mK_2, the graph $n(mK_2)$ admits at least $|\mathscr{S}_n|^m$ nonisomorphic SEM labelings. By Corollary 6.6, we also know that there are at least $N(m)|\mathscr{S}_n|^m$ nonisomorphic SEM labelings of $n(mK_2)$. Interchanging the roles of m and n, the result follows.   □

Let $B(r) = \{(m,n) \in \mathbb{N} \times \mathbb{N} : mn = r\}$. The previous result generalizes into the next corollary.

Corollary 6.8 ([16])   *Let l be an odd positive integer. Then, the graph lK_2 admits at least $\max_{(m,n)\in B(l)}\{N(n)|\mathscr{S}_m|^n\}$ (S)(S)EM labelings.*

Example 6.7 ([16]) By combining all SSEM labelings of $3K_2$ with all elements of $\mathscr{S}_3$, we obtain 16 nonisomorphic SSEM labelings of $9K_2$. Assume that V_1 and V_2 are the stable sets of $9K_2$. We enumerate all these labelings in Table 6.1. Let the left column be the labels of V_1. Then, each remaining column corresponds to the labelings of the stable set V_2.

Using Theorem 6.3, we can obtain lower bounds for the number of nonisomorphic (S)(S)EM labelings of odd copies of a bipartite (S)(S)EM graph.

Table 6.1 Nonisomorphic SSEM labelings of $9K_2$

1	14	15	14	14	14	15	15	15	17	17	17	18	17	18	18	18
2	15	13	15	15	15	13	13	13	18	18	18	16	18	16	16	16
3	13	14	13	13	13	14	14	14	16	16	16	17	16	17	17	17
4	17	17	18	17	18	17	18	18	11	12	11	11	12	12	11	12
5	18	18	16	18	16	18	16	16	12	10	12	12	10	10	12	10
6	16	16	17	16	17	16	17	17	10	11	10	10	11	11	10	11
7	11	11	11	12	12	12	11	12	14	14	15	14	15	14	15	15
8	12	12	12	10	10	10	12	10	15	15	13	15	13	15	13	13
9	10	10	10	11	11	11	10	11	13	13	14	13	14	13	14	14

Corollary 6.9 ([16]) *Let G be a bipartite (S)(S)EM graph and let n be an odd positive integer. Then every (S)(S)SEM labeling of G generates at least $|\mathscr{S}_n|$ nonisomorphic (S)(S)EM labelings of nG.*

Proof By Theorems 6.3 and 6.5, and by Corollary 6.1 the result comes easily. We observe that changing the function $h : E(\overrightarrow{G}) \to \mathscr{S}_n$ all the labelings that we obtain are nonisomorphic since all copies of one of the stable sets have the same labels, for every h.   □

Exercise 6.2 Denote by $\sharp(G)$ the number of super edge-magic labelings of a graph G.

(a) Show that

$$\sharp(P_4 \cup (2n-1)K_2) = \begin{cases} (3n+2)\sharp((2n+1)K_2), & n \text{ even}, \\ (3n+1)\sharp((2n+1)K_2), & n \text{ odd}. \end{cases}$$

(b) Denote by C_{2n+1}^C the set of all cycles of order $2n+1$ with a chord. Show that

$$\sum_{H \in C_{2n+1}^C} \sharp(H) = \begin{cases} n\sharp(C_{2n+1}), & n \text{ even}, \\ (n-1)\sharp(C_{2n+1}), & n \text{ odd}. \end{cases}$$

6.5   Relations Obtained from the $\otimes_h$-Product

This section contains mainly applications of the $\otimes_h$-product to labelings involving as a second factor the family $\mathscr{S}_n^k$ that has been introduced in Sect. 6.3. The results can be found in [31] and almost all correspond to generalizations of previous results found in [26, 30].

The concepts of edge bi-magic and super edge bi-magic labelings were first introduced by Babujee in [2, 3].

Definition 6.3 Let G be a (p,q)-graph and let $f : V(G) \cup E(G) \to [1, p+q]$ be a bijective function such that $f(u) + f(uv) + f(v) \in \{k_1, k_2\} \subset \mathbb{N}$, for all $uv \in E(G)$. Then f is called an *edge bi-magic labeling* of G and G is called an *edge bi-magic graph*. The integers k_1, k_2 are called the *valences* of f. An edge bi-magic labeling f of G which verifies the extra condition $f(V(G)) = [1, p]$ is called *super edge bi-magic* and G is called a *super edge bi-magic graph*.

With a similar proof as the one of Theorem 6.5 we can prove the following.

Theorem 6.14 ([31]) *Let D be a (super) edge bi-magic digraph and let $h : E(D) \to \mathscr{S}_n^k$ be any function. Then the graph und$(D \otimes_h \mathscr{S}_n^k)$ is (super) edge bi-magic.*

Exercise 6.3 Prove Theorem 6.14.

The next theorem is similar to Theorem 6.5 for harmonious graphs.

Theorem 6.15 ([31]) *Let D be a harmonious (p,q)-digraph with $p \leq q$ and let $h : E(D) \longrightarrow \mathscr{S}_n^k$ be any function. Then $und(D \otimes_h \mathscr{S}_n^k)$ is harmonious.*

Proof We rename the vertices of D after the labels of a harmonious labeling. We consider a slight modification of the labels introduced in the proof of Theorem 6.5: if $(i,j) \in V(D \otimes_h \mathscr{S}_n^k)$ we assign to the vertex the label $ni + j - 1 \pmod{nq}$.

Given an arc $((i,j),(i',j')) \in E(D \otimes_h \mathscr{S}_n^k)$, coming from an arc $e = (i,i') \in E(D)$ and an arc $(j,j') \in E(h(i,i'))$, the induced arc label is equal to:

$$n(i + i') + j + j' - 2 \pmod{nq}. \tag{6.4}$$

Since D is harmonious, the set $\{i + i' \pmod{q} : (i,i') \in E(D)\}$ covers all elements in $\mathbb{Z}_q$. Whereas since each element F of $\mathscr{S}_n^k$ is labeled with a super edge-magic labeling with minimum sum of the adjacent vertex labels equal to k, we have

$$\{(j + j') : (j,j') \in E(\Gamma)\} = [k, k + n - 1].$$

Thus, it is easy to check that the set of arc labels covers all the elements in $\mathbb{Z}_{nq}$, and the result follows.   □

The authors of [26, 31] showed that the set of labelings in which we can use the $\otimes_h$-product to generate new families of labeled graphs includes sequential and partitional labelings.

Theorem 6.16 ([31]) *Let D be a sequential digraph and let $h : E(D) \longrightarrow \mathscr{S}_n^k$ be any function. Then $und(D \otimes_h \mathscr{S}_n^k)$ is sequential.*

Proof We rename the vertices of D after the labels of a sequential labeling. Similarly to the proof of Theorem 6.15, if $(i,j) \in V(D \otimes_h \mathscr{S}_n)$ we assign to the vertex the label $ni + j - 1$.

Given an arc $((i,j)(i',j')) \in E(D \otimes_h \mathscr{S}_n)$, coming from an arc $e = (i,i') \in E(D)$ and an arc $(j,j') \in E(h(i,i'))$, the induced arc label is equal to: $n(i + i') + j + j' - 2$.

Since D is sequential, the set $\{i + i' : (i,i') \in E(D)\}$ covers all elements in $[m, m + |E(D)| - 1]$, for some positive integer m. Whereas since each element F of $\mathscr{S}_n^k$ is labeled with a super edge-magic labeling with minimum induced edge sum being equal to k, we have $\{(j + j') : (j,j') \in E(F)\} = [k, k + n - 1]$.

Thus, an easy check shows that the set of arc labels covers all elements in

$$k - 2 + [nm, n(m + |E(D)|) - 1] = [m', m' + n|E(D)| - 1],$$

where $m' = k + mn - 2$.   □

A particular case of a sequential labeling, namely partitional labeling, was introduced by Ichishima and Oshima [25].

Definition 6.4 Let G be a bipartite graph of size $2t + s$ with stable sets U and V of the same cardinality s. We say that a sequential labeling of G is *partitional* if: (a) $f(u) \leq t + s - 1$ for each $u \in U$ and $f(v) \geq t - s$ for each $v \in V$, (b) there is a positive integer m such that the induced edge labels are partitioned into three sets: $[m, m+t-1] \cup [m+t, m+t+s-1] \cup [m+t+s, m+2t+s-1]$, and there is an involution π (automorphism) of G such that (i) π exchanges U and V, (ii) $u\pi(u) \in E(G)$, for all $u \in U$, and (iii) $\{f(u) + f(\pi(u)) : u \in U\} = [m + t, m + t + s - 1]$. A graph that admits a partitional labeling is called a *partitional graph*.

We leave as an exercise the proof of the next theorem. The proof is similar to the one of Theorem 3.2 in [26], but taking $m' = mn + k - 2$.

Theorem 6.17 ([31]) *Let G be a partitional graph and let $h : E(\vec{G}) \longrightarrow \mathscr{S}_n^k$ be any function, where $\vec{G}$ is the digraph obtained by orienting all edges from one stable set to the other one. Then $und(\vec{G} \otimes_h \mathscr{S}_n^k)$ is partitional.*

The product can be applied also to obtain families of cordial graphs [12].

Theorem 6.18 ([31]) *Let D be a cordial digraph and let $h : E(D) \longrightarrow \mathscr{S}_n^k$ be any function. Then $und(D \otimes_h \mathscr{S}_n^k)$ is cordial.*

Proof We rename the vertices of each element of $\mathscr{S}_n^k$ after the labels of a super edge-magic labeling. Let f be a cordial labeling of D.

We will prove that, the labeling on $V(D \otimes_h \mathscr{S}_n^k)$ defined by

$$l(u, j) = \alpha \in \{0, 1\}, \quad \text{where} \quad \alpha \equiv f(u) + j - 1 \pmod{2} \tag{6.5}$$

is a cordial labeling of $D \otimes_h \mathscr{S}_n^k$. Let $V(D) = V_0 \cup V_1$, where $V_i = f^{-1}(i)$, for $i = 0, 1$. Since $1 - f$ is also a cordial labeling of D, we can assume that $|V_0| \leq |V_1| \leq |V_0| + 1$. Also consider the partition $[1, n] = I_0 \cup I_1$, where $I_1 = [1, n] \cap \{1, 3, 5, \ldots\}$ (note that $|I_0| \leq |I_1| \leq |I_0| + 1$).

Let us check the condition on the vertices. By definition of l, we have $l^{-1}(0) = V_0 \times I_1 \cup V_1 \times I_0$ and $l^{-1}(1) = V_0 \times I_0 \cup V_1 \times I_1$. Thus,

(a)   If $|V_0| = |V_1| = r$ and

    (a.1) $|I_0| = |I_1| = s$. Then, $|l^{-1}(0)| = 2rs = |l^{-1}(1)|$.
    (a.2) $|I_0| = |I_1| - 1 = s$. Then, $|l^{-1}(0)| = 2rs + r = |l^{-1}(1)|$.

(b)   If $|V_0| = |V_1| - 1 = r$ and

    (b.1) $|I_0| = |I_1| = s$. Then, $|l^{-1}(0)| = 2rs + s = |l^{-1}(1)|$.
    (b.2) $|I_0| = |I_1| - 1 = s$. Then, $|l^{-1}(0)| = 2rs + r + s = |l^{-1}(1)| - 1$.

Now, let us check the condition on the arcs. Let f_a be the labeling induced on the arcs of D by f and let $E(D) = E_0 \cup E_1$, where $E_i = f_a^{-1}(i)$, for $i = 0, 1$. Also consider the partition $[k, k+n-1] = J_0 \cup J_1$, where $J_1 = [k, k+n-1] \cap \{1, 3, 5, \ldots\}$.

Notice that, by equality (6.5), the labeling l_a induced on the arcs assigns to $((u, i), (v, j))$ the label $|\beta|$, where $\beta \in \{0, 1\}$ and $\beta \equiv f(u) - f(v) + (i - j) \pmod{2}$.

But, since $i - j$ is of the same parity as $i + j$, we have $l_a^{-1}(0) = E_0 \times J_0 \cup E_1 \times J_1$ and $l_a^{-1}(1) = E_0 \times J_1 \cup E_1 \times J_0$. Thus,

(a)   If $|E_0| = |E_1| = r$ and

    (a.1) $|J_0| = |J_1| = s$. Then, $|l_a^{-1}(0)| = 2rs = |l_a^{-1}(1)|$.

    (a.2) $|J_0| = |J_1| - 1 = s$. Then, $|l_a^{-1}(0)| = 2rs + r = |l_a^{-1}(1)|$.

    (a.3) $|J_0| - 1 = |J_1| = s$. Then, $|l_a^{-1}(0)| = 2rs + r = |l_a^{-1}(1)|$.

(b)   If $|E_0| = |E_1| - 1 = r$ and

    (b.1) $|J_0| = |J_1| = s$. Then, $|l_a^{-1}(0)| = 2rs + s = |l_a^{-1}(1)|$.

    (b.2) $|J_0| = |J_1| - 1 = s$. Then, $|l_a^{-1}(0)| = 2rs + r + s = |l_a^{-1}(1)| - 1$.

    (b.3) $|J_0| - 1 = |J_1| = s$. Then, $|l_a^{-1}(0)| = 2rs + r + s = |l_a^{-1}(1)| - 1$.

(c)   If $|E_0| - 1 = |E_1| = r$ and

    (c.1) $|J_0| = |J_1| = s$. Then, $|l_a^{-1}(0)| = 2rs + s = |l_a^{-1}(1)|$.

    (c.2) $|J_0| = |J_1| - 1 = s$. Then, $|l_a^{-1}(0)| = 2rs + r + s = |l_a^{-1}(1)| - 1$.

    (c.3) $|J_0| - 1 = |J_1| = s$. Then, $|l_a^{-1}(0)| - 1 = 2rs + r + s = |l_a^{-1}(1)|$.

□

6.5.1 Labelings Involving Differences

Bloom and Ruiz introduced in [11] a generalization of graceful labelings, that they called k-equitable labelings.

Definition 6.5 Let G be a (p, q)-graph and let $g : V(G) \longrightarrow \mathbb{Z}$ be an injective function with the property that the new function $h : E(G) \longrightarrow \mathbb{N}$ defined by the rule $h(uv) = |g(u) - g(v)|$, for all $uv \in E(G)$ assigns the same integer to exactly k edges. Then g is said to be a k-*equitable labeling* and G a k-*equitable graph*. A k-equitable labeling is said to be *optimal* [11], when g assigns all the elements of the set $[1, p]$ to the elements of $V(G)$.

López et al. [30, 31] used the $\otimes_h$-product in order to construct k-equitable labelings of new families of graphs. The input elements in [30] were k-equitable digraphs and the family $\mathscr{S}_n$, but instead of applying the product directly, the authors introduced what they called the rotation of a super edge-magic digraph. In [31], the authors extended this process to the more general family $\mathscr{S}_n^{\sigma}$. However, the restriction that all the minimum induced sums σ (see Lemma 2.1) have to be equal to $(n + 3)/2$ should be assumed. This implies that only families with magic sum that coincides with the same magic sum of a super edge-magic cycle of length n are accepted.

We start by recalling the concept that was introduced in [30] and showing which results can be applied to the more general family of $\mathscr{S}_n^{(n+3)/2}$.

Let $M = (a_{i,j})$ be a square matrix of order n. The matrix $(a_{i,j}^R)$ is the *rotation of the matrix M*, denoted by M^R, when $a_{i,j}^R = a_{n+1-j,i}$. Graphically this corresponds to a rotation of the matrix by $\pi/2$ radiants clockwise. (By rotating $3\pi/2$ radiants clockwise the matrix M we obtain the matrix M^{3R}, which has the same properties that we look for in M^R.)

Lemma 6.6 ([31]) *Let $F \in \mathscr{S}_n^{(n+3)/2}$ and let $A = (a_{i,j})$ be its adjacency matrix. If $a_{i,j}^R = 1$, then*

$$|i - j| \le \frac{n-1}{2}.$$

Proof Since the minimum sum of the adjacent vertices in $\mathscr{S}_n^{(n+3)/2}$ is $(n+3)/2$, if $A = (a_{i,j})$ is the adjacency matrix of $F \in \mathscr{S}_n^{(n+3)/2}$ and $a_{i,j} = 1$, we have $(n+3)/2 \le i+j \le (3n+1)/2$. Hence, since $a_{i,j}^R = a_{n+1-j,i}$, if $a_{i,j}^R = 1$ it follows that $(n+3)/2 \le n+1-j+i \le (3n+1)/2$. Therefore, $-(n-1)/2 \le i-j \le (n-1)/2$ and we obtain the result. $\square$

A digraph S is said to be a *rotation super edge-magic digraph of order n and minimum sum k*, if its adjacency matrix is the rotation of the adjacency matrix of an element in $\mathscr{S}_n^k$. We denote by $\mathscr{RS}_n^k$ the set of all digraphs that are rotation super edge-magic digraphs of order n and minimum sum k.

Example 6.8 Let S be the digraph of Fig. 6.7b. Then S is a rotation super edge-magic digraph of order 3 and minimum sum 3. If we consider the adjacency matrix of S, we have $A(S) = \begin{pmatrix} 1 & 0 & 0 \\ 0 & 0 & 1 \\ 0 & 1 & 0 \end{pmatrix}$. This matrix is the rotation matrix of $A(D) = \begin{pmatrix} 0 & 1 & 0 \\ 0 & 0 & 1 \\ 1 & 0 & 0 \end{pmatrix}$, where D is the super edge-magic digraph that appears on Fig. 6.7a.

The following corollary is an easy observation.

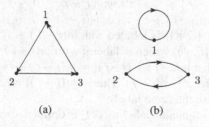

(a)　　　　　(b)

Fig. 6.7 (a) A super edge-magic digraph and (b) the digraph obtained by rotating its adjacency matrix

Corollary 6.10 ([31]) *Let S be a digraph in $\mathscr{RS}_n^{(n+3)/2}$ and let k be any integer. If $|k| \leq (n-1)/2$, then there exists a unique arc $(i,j) \in E(S)$ such that $i - j = k$.*

Assume that D is a k-equitable digraph where the vertices are identified by the labels of a k-equitable labeling of D. Let $h : E(D) \to \mathscr{RS}_n^{(n+3)/2}$ be any function. If we consider the induced labeling on $V(D \otimes_h \mathscr{RS}_n^{(n+3)/2})$ that assigns the label $n(i-1) + j$ to the vertex (i,j), then all labels are distinct and, in case the labeling of D is optimal, all elements in $[1, n|V(D)|]$ are used. Moreover, by the $\otimes_h$-product's definition, $|n(i-i') + (j-j')|$ is an induced arc label if and only if $(i, i') \in E(D)$ and $(j, j') \in E(h(i, i'))$.

Lemma 6.7 ([31]) *Let D be a k-equitable digraph, and let $((i,j),(i',j'))$, $((r,s),(r',s'))$ be two arcs of $D \otimes_h \mathscr{RS}_n^{(n+3)/2}$, for some function $h : E(D) \to \mathscr{RS}_n^{(n+3)/2}$. If $|n(i-i') + (j-j')| = |n(r-r') + (s-s')|$, then $|i-i'| = |r-r'|$ and $|s-s'| = |j-j'|$.*

Proof Note that the equality $n(i-i') + (j-j') = \pm(n(r-r') + (s-s'))$ implies that there exists $\alpha \in \mathbb{Z}$ such that $|\alpha n| = |\pm (s-s') - (j-j')|$. By Lemma 6.6, we get $|\alpha n| \leq n-1$. Thus, we obtain $\alpha = 0$ and therefore, $|j-j'| = |s-s'|$ and $|i-i'| = |r-r'|$.   □

Now, we present an application of the $\otimes_h$-product for k-equitable digraphs.

Theorem 6.19 ([31]) *Let D be an (optimal) k-equitable digraph and let $h : E(D) \to \mathscr{RS}_n^{(n+3)/2}$ be any function. Then $D \otimes_h \mathscr{RS}_n^{(n+3)/2}$ is (optimal) k-equitable.*

Proof Assume that $|n(i-i')+(j-j')|$ is an arc label induced by a k-equitable labeling of D. There exist exactly k arcs in D, (i_l, i'_l), $1 \leq l \leq k$ such that $|i_l - i'_l| = |i - i'|$. Thus, $|n(i_l - i'_l)| = |n(i - i')|$ and by Lemma 6.6 we obtain

$$|n(i_l - i'_l)| - \frac{n-1}{2} \leq |n(i-i') + (j-j')| \leq |n(i_l - i'_l)| + \frac{n-1}{2}.$$

Hence, we have $||n(i-i') + (j-j')| - |n(i_l - i'_l)|| \leq (n-1)/2$ and by Corollary 6.10 there exist two different arcs $(r, r'), (s, s') \in E(h(i_l, i'_l))$ such that $||n(i - i') + (j - j')| - |n(i_l - i'_l)|| = |r - r'| = |s - s'|$ with $r - r' \leq 0 \leq s - s'$. Therefore, either $|n(i-i') + (j-j')| = |n(i_l - i'_l) + r - r'|$ or $|n(i-i') + (j-j')| = |n(i_l - i'_l) + s - s'|$. In the first case, $((i_l, r), (i'_l, r'))$ is labeled with $|n(i - i') + (j - j')|$, whereas in the second case, is $((i_l, s), (i'_l, s'))$ which is labeled with $|n(i - i') + (j - j')|$.

Moreover, assume that $|n(i-i') + (j-j')| = |n(r-r') + (s-s')|$. By Lemma 6.7, $|i - i'| = |r - r'|$ and $|s - s'| = |j - j'|$. That is, $|n(i - i')| = |n(r - r')|$ and we only have k-possible arcs with the same label.

In particular, if the k-equitable labeling of D is optimal, then the induced labeling on $D \otimes_h \mathscr{RS}_n^{(n+3)/2}$ is also optimal.   □

Example 6.9 Figure 6.8 shows an optimal 2-equitable labeling of the digraph D obtained by orientating the edges of C_6, from one stable set to the other. The digraph

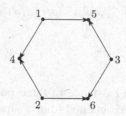

Fig. 6.8 An optimal 2-equitable labeling of D

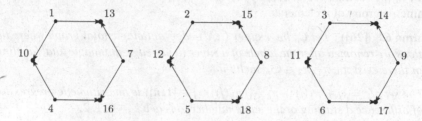

Fig. 6.9 The induced optimal 2-equitable labeling of $D \otimes S$

$D \otimes S$ is also 2-equitable, where S is the rotation super edge-magic of order 3 and minimum sum 3 that appears in Fig. 6.7b. An optimal 2-equitable labeling of $D \otimes S$, induced by the labelings of D and S, is shown in Fig. 6.9.

6.5.2 (Super) (a, d)-Edge-Antimagic Total Labelings

Recently, a lot of interest has emerged in relation to labelings of the antimagic type. A good proof for this is the book [4], and, for instance, the following papers [5–9, 13, 36] that have recently appeared in the literature. In this section we concentrate on (a, d)-edge-antimagic total $((a, d)$-EAT, for short) labelings that were introduced by Simanjuntak et al. in [36].

Dafik et al. formulated in [13] the following question: "if a graph G is super (a, d)-EAT, is the disjoint union of multiple copies of the graph G (a, d)-EAT as well?" They answered this question when the graph G is either a cycle or a path.

It was first proved in an unpublished paper by Kotzig [27] (see also [33]) and later, independently and unaware of Kotzig's work, it was reproved by Figueroa et al. [15] that if G is a tripartite graph which admits a (super) $(a, 0)$-EAT labeling and n is odd then the graph nG also admits a (super) $(a, 0)$-EAT labeling. Following the same line of research, Bača et al. [9] have shown that if G is a tripartite graph which admits a (super) $(a, 2)$-EAT then the graph nG also admits a (super) $(a, 2)$-EAT labeling. The main goals in this section are to generalize the results established so far to the case when $d = 1$ and to introduce new proofs of these results based on the Kronecker product of digraphs, that we feel that give more inside to the problem than the proofs known so far.

Bača et al. proved in [5] the following result.

Theorem 6.20 ([5]) *The cycle C_n has a super (a,d)-edge-antimagic total labeling if and only if either*

(i) $d = 0, 2$ and n is odd, $n \geq 3$, or
(ii) $d = 1$ and $n \geq 3$.

The next lemma shows the existence of three permutations in the symmetric group of n elements that can be obtained from a super (a, d)-edge-antimagic total labeling of the cycle. These permutations will be used in the proof of the main result of the section. We denote by $+_k$ the sum of integers (mod k) and by $\mathfrak{S}_n$ the symmetric group of n elements.

Lemma 6.8 ([26]) *Let C_n be a super (a, d)-edge-antimagic total graph where the vertices are renamed after the labels of a super (a, d)-edge-antimagic total labeling. Then there exist $\pi_0, \pi_1, \pi_2 \in \mathfrak{S}_n$ such that:*

- *The set $\mathscr{A}_k = \{j + \pi_k(j) + \pi_{k+_31}(\pi_k(j)) : j \in [1, n]\}$ is an arithmetic progression of difference d starting at the same number for each $k = 0, 1, 2$.*
- *$\pi_2 \circ \pi_1 \circ \pi_0 = id$,*

where id denotes the identity permutation.

Proof We rename the vertices and the arcs of C_n^+ after the labels of a super (a, d)-EAT labeling. Let e_u be the label assigned to the arc (u, v). We define the following permutations:

$$\pi_0(u) = e_u - n, \quad \pi_1(e_u - n) = v \quad \text{and} \quad \pi_2(v) = u.$$

Clearly, $\mathscr{A}_0 = \{u + \pi_0(u) + \pi_1(\pi_0(u)) = u + e_u - n + v : (u, v) \in E(C_n^+)\}$ defines an arithmetic progression starting at $a - n$ and with difference d. The same works for $\mathscr{A}_1$ and $\mathscr{A}_2$. $\square$

Example 6.10 ([26]) Let us see an example of the previous lemma for $n = 5$. From the $(10, 2)$-edge-antimagic total labeling of C_5^+ that appears in Fig. 6.10a, we obtain the three permutations that appear in Fig. 6.11.

Next we prove the following result found in [9, 15, 27, 33] using a different argument. It sheds some new light on the reasons why the theorem is true. Furthermore the proof allows us to construct many different (a, d)-EAT labelings of the resulting graph.

Theorem 6.21 ([26]) *If G is a (super) (a, d)-edge-antimagic total tripartite graph, then nG is (super) (a', d)-EAT, where $n \geq 3$,*

(i) $d = 0, 2$ and n is odd, or
(ii) $d = 1$.

Proof For the values of n considered in the statement of Theorem 6.21, we know by Theorem 6.20 that the cycle C_n admits a super (a, d)-edge-antimagic total labeling. Thus by Lemma 6.8 there exist three permutations π_0, π_1, and π_2 in $\mathfrak{S}_n$ such that

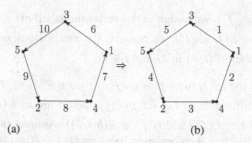

Fig. 6.10 (a) A super $(10, 2)$-edge-antimagic total labeling of C_5^+ and (b) an induced labeling of C_5^+

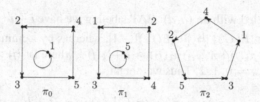

Fig. 6.11 The three permutations coming from the labeling of Fig. 6.10

the set

$$\mathscr{A}_k = \{j + \pi_k(j) + \pi_{k+_3 1}(\pi_k(j)) : j \in [1, n]\}$$

is an arithmetic progression with difference d for each $k = 0, 1, 2$. Let us denote by F_k the 1-regular digraphs whose adjacency matrices correspond to the graphic representation of each of the permutations π_k, for $k \in [0, 2]$. We let $\Gamma = \{F_0, F_1, F_2\}$.

We rename the vertices and the edges of G after the labels of a super (a, d)-EAT labeling. Let V_0, V_1, and V_2 be the stable sets of the graph G and let us denote by $\overrightarrow{G}$ the digraph obtained from G by orienting each edge from V_k to $V_{k+_3 1}$. Let $h : E(\overrightarrow{G}) \longrightarrow \Gamma$ be the function defined by:

$$h((u, v)) = F_k \quad \text{if} \quad u \in V_k.$$

We show that $und(\overrightarrow{G} \otimes_h \Gamma) = nG$.

For each $j \in [0, n-1]$ the subdigraph of $\overrightarrow{G} \otimes_h \Gamma$ induced by

$$(V_0 \times \{j\}) \cup (V_1 \times \{\pi_0(j)\}) \cup (V_2 \times \{\pi_1(\pi_0(j))\})$$

is isomorphic to $\overrightarrow{G}$. This is clear since, by Lemma 6.8, we know that $\pi_2 \circ \pi_1 \circ \pi_0 = id$. Next we claim that the graph nG is (super) (a', d)-edge-antimagic total. To prove this, we only have to consider the following induced labeling f:

1. If $(i,j) \in V(\vec{G} \otimes_h \Gamma)$, we assign to the vertex the label: $n(i-1)+j$.
2. If $((i,j),(i',j')) \in E(\vec{G} \otimes_h \Gamma)$, we assign to the arc the label: $n(e-1)+\pi_{k+_31}(j')$,
   where e is the label of (i,i') in $\vec{G}$ and $i \in V_k$.

Let us see now that the set $\{f(u)+f(uv)+f(v) : uv \in E(\vec{G} \otimes_h \Gamma)\}$ is an arithmetic progression with difference d. Let $((i,j)(i',j'))$ be an arc in $E(\vec{G} \otimes_h \Gamma)$ coming from arcs $e = (i,i') \in E(\vec{G})$ and $(j,j') \in E(h(i,i'))$. Assume that $i \in V_k$, thus by definition $j' = \pi_k(j)$. Then the corresponding sum $f(u)+f(uv)+f(v)$ is equal to:

$$n(i+i'+e-3)+j+\pi_k(j)+\pi_{k+_31}(\pi_k(j)). \qquad (6.6)$$

Since $\vec{G}$ is labeled with an (a,d)- EAT labeling we have $i+i'+e = a+\mu(e)d$ where $\{\mu(e) : e \in E(\vec{G})\} = [0,|E(\vec{G})|-1]$, whereas by Lemma 6.8 there exists $b \in \mathbb{Z}$ such that $j+\pi_k(j)+\pi_{k+_31}(\pi_k(j)) = b+\nu_k(j)d$, where $\{\nu_k(j) : j \in [0,n-1]\} = [0,n-1]$ for each $k = 0,1,2$. Thus we obtain

$$n(i+i'+e-3)+j+\pi_k(j)+\pi_{k+_31}(\pi_k(j)) = n(a-3)+b+(n\mu(e)+\nu_k(j))d.$$

Therefore, the set of sum labels of $\vec{G} \otimes_h \Gamma$ is an arithmetic progression starting at $n(a-3)+b$ and with difference d.

Notice that, if the digraph $\vec{G}$ is super EAT then the vertices of $\vec{G} \otimes_h \Gamma$ receive the smallest labels. $\qquad\qquad \square$

As a corollary we obtain a result that is contained in [13].

Corollary 6.11 ([26]) *Let* $m,n \geq 3$. *The graph* nC_m *has a super* (a,d)-*edge-antimagic total labeling when*

(i) $d = 0,2$ *and, m and n are odd, or*
(ii) $d = 1$.

Bača et al. showed in [6] that P_n, $n \geq 2$, has a super (a,d)-EAT labeling if and only if $d \in \{0,1,2,3\}$. Using this result and Theorem 6.21 we obtain the next result that also appears in [13].

Corollary 6.12 ([13, 26]) *Let* $m \geq 2$ *and* $n \geq 3$. *The graph* nP_m *has a super* (a,d)-*edge-antimagic total labeling when*

(i) $d = 0,2$ *and n is odd, or*
(ii) $d = 1$.

A summary of the relations established in this section can be visualized in Fig. 6.12.

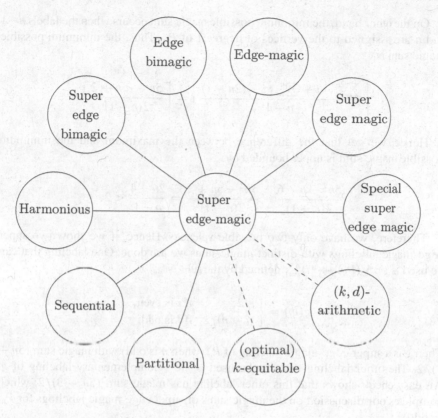

Fig. 6.12 Relations obtained from the $\otimes_h$-product

6.6 Further Applications

We recall that in Corollary 3.6, we have established that for all $n \in \mathbb{N} \setminus \{1\}$, there exists a super edge-magic graph G such that $|\sigma_1 - \sigma_2| \geq n - 1$, where σ_1 and σ_2 are the only two possible distinct magic sums of G. The path P_n, n odd, is an example of the contrary situation. That is to say, if $\{\sigma_1 < \sigma_2 < \ldots < \sigma_l\}$ is the set of all possible magic sums for super edge-magic labelings of P_n (n odd), then $\sigma_{i+1} - \sigma_i = 1$ for all $i \in [1, l-1]$. Next, we show this fact.

Consider the path P_n when n is odd. Then the maximum possible magic sum occurs when labels 1 and 2 are assigned to the vertices of degree 1 of P_n. Thus, the maximum possible magic sum is

$$\left\lfloor \frac{\sum_{i=1}^{2n-1} i + \sum_{i=1}^{n} i - 3}{n-1} \right\rfloor = \left\lfloor \frac{5n^2 - n - 6}{2(n-1)} \right\rfloor.$$

On the other hand, the minimum possible magic sum occurs when the labels $n-1$ and n are assigned to the vertices of degree 1 of P_n. Thus, the minimum possible magic sum is

$$\left\lceil \frac{\sum_{i=1}^{2n-1} i + \sum_{i=1}^{n} i - (2n-1)}{n-1} \right\rceil = \left\lceil \frac{5n^2 - 5n + 2}{2(n-1)} \right\rceil.$$

Hence, we get that the difference between the maximum and the minimum possible magic sum is upper bounded by:

$$\frac{5n^2 - n - 6}{2(n-1)} - \frac{5n^2 - 5n + 2}{2(n-1)} = \frac{2n-4}{n-1} < 2.$$

Therefore, we have only two possible valences. Hence, if we show two super edge-magic labelings with distinct magic sums we are done. One labeling that can be used is $g : V(P_n) \to \{i\}_{i=1}^{n}$ defined by the rule

$$g(v_i) = \begin{cases} i/2, & \text{if } i \text{ is even,} \\ (i+n)/2, & \text{if } i \text{ is odd.} \end{cases}$$

Then g is a super edge-magic labeling of P_n (when n is odd) with magic sum $(5n + 1)/2$. The other labeling that can be used is the s-complementary labeling of g. An easy check shows that this other labeling has magic sum $(5n + 3)/2$, which completes our discussion on the magic sums of super edge-magic labelings for P_n, (n odd).

Exercise 6.4 What can be said about the set of possible valences of the family of paths P_n, when n is even?

All cycles are edge-magic [28]. However, a cycle C_n is super edge-magic if and only if n is odd [14]. Godbold and Slater introduced in [18] the following conjecture.

Conjecture 6.1 ([18]) For $n = 2t + 1 \geq 7$ and $5t + 4 \leq j \leq 7t + 5$ there is an edge-magic labeling of C_n with magic sum $k = j$. For $n = 2t \geq 4$ and $5t + 2 \leq j \leq 7t + 1$ there is an edge-magic labeling of C_n with magic sum $k = j$.

This is an old problem that appeared in 1998 and that has remained unsolved for 18 years. Very little progress has been made towards a solution of it since then. In fact, for many years only four magic sums have been known for C_n, except for small values of n, where the problem has been treated using computers (see [10]). It has not been until 2009 that a paper has appeared [34], in which McQuillan proves the next result.

Theorem 6.22 ([34]) *Assume that C_k has an edge-magic labeling with magic sum g. Let $n = 2m + 1$ be an odd integer. Then, C_{nk} has an edge-magic labeling with magic sum $ng - 3m$ and an edge-magic labeling with magic sum $6km + g$.*

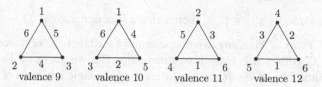

Fig. 6.13 Edge-magic labelings of C_3

A similar result, but this time using the $\otimes_h$-product can be obtained. Since the main goal of this chapter is to emphasize the power of the $\otimes_h$-product, we will concentrate on this technique.

The following two results appear in [33].

Theorem 6.23 ([33]) *Every odd cycle C_n has an edge-magic labeling with magic sum $3n + 1$ and an edge-magic labeling with magic sum $3n + 2$.*

Theorem 6.24 ([33]) *Every even cycle C_n has an edge-magic labeling with magic sum $(5n + 4)/2$.*

The cycle C_3 admits four edge-magic labelings with different magic sums. See Fig. 6.13.

Lemma 6.9 *Let $n \geq 3$ be an odd integer and suppose that $m \geq 3$ is an integer such that either m is odd or $m \geq n$. Then there exists a function $h : E(C_m^+) \to \{C_n^+, C_n^-\}$ such that*

$$C_m^+ \otimes_h \{C_n^+, C_n^-\} \cong C_{mn}^+.$$

Proof We have $\langle 1 \rangle = \mathbb{Z}_n$ and, since n is odd, the congruence relation $m - 2r \equiv 1 \ (mod \ n)$ can be solved, with $0 < r < m$. Therefore, inheriting the notation of Theorem 6.4, we get the desired result.

Now, we would like to end this chapter by proving the following result.

Theorem 6.25 ([32]) *For all $n_0 \in \mathbb{N}$, there exists $n \in \mathbb{N}$, such that the cycle C_n admits at least n_0 edge-magic labelings with at least n_0 mutually distinct magic sums.*

Proof We already know that C_3 admits 4 edge-magic labelings with 4 consecutive edge-magic magic sums (notice that the labeling corresponding to magic sum 9 is super edge-magic). Call these labelings l_1, l_2, l_3, l_4, where the magic sum of l_i is less than the magic sum of l_j if and only if $i < j$ ($i, j \in [1, 4]$). Denote by $C_3^{l_i}$ the copy of C_3, where each vertex takes the name of the label that l_i has assigned to it. Also let $\overrightarrow{C_3^{l_i}}$ be the digraph obtained from $C_3^{l_i}$ with the edges oriented cyclically. Let $\Gamma = \{C_3^+, C_3^-\}$, where we assume that the vertices of C_3 are labeled in a super edge-magic way.

By Lemma 6.9, for all $i \in [1, 4]$ there exists a function $h_i : E(\overrightarrow{C_3^{l_i}}) \to \Gamma$ such that und$(\overrightarrow{C_3^{l_i}} \otimes_{h_i} \Gamma) \cong C_9$. Also, any two magic sums of the labelings obtained for $\overrightarrow{C_3^{l_i}} \otimes_{h_i} \Gamma$ differ, by Corollary 6.2, by at least three units. But we know by Theorem 6.23 that magic sums 28 and 29 appear for different edge-magic labelings of C_9. Hence, the cycle C_9 admits at least 5 edge-magic labelings with 5 different magic sums. Let the labelings that provide these magic sums be l_i^1, where the magic sum of l_i^1 is less than the magic sum of l_j^1 if and only if $i < j$ $(i, j \in [1, 5])$.

If we repeat the process with $\overrightarrow{C_9^{l_i^1}} \otimes_{h_i^1} \Gamma$, where $h_i^1 : E(\overrightarrow{C_9^{l_i^1}}) \to \Gamma$ is a function as in Lemma 6.9, we obtain 5 edge-magic labelings of C_{27} with 5 different magic sums. But, again by Corollary 6.2, either magic sum 82 or magic sum 83 does not appear among these 5 magic sums, since among these 5 magic sums no two magic sums are consecutive. But we know by Theorem 6.23 that these two magic sums, 82 and 83, appear for an edge-magic labeling of C_{27}. Hence, there are at least 6 magic sums for edge-magic labelings of C_{27}.

Repeating this process inductively, we obtain that each cycle of order 3^α admits at least $3 + \alpha$ edge-magic labelings with at least $3 + \alpha$ mutually different magic sums. Therefore, we get the desired result. $\square$

Notice that, using a similar idea to the one in the proof of Theorem 6.19, we can obtain the next theorem.

Theorem 6.26 ([32]) *Let $n = p_1^{\alpha_1} p_2^{\alpha_2} \cdots p_k^{\alpha_k}$ be the unique prime factorization (up to ordering) of an odd number n. Then C_n admits at least $1 + \sum_{i=1}^{k} \alpha_i$ edge-magic labelings with at least $1 + \sum_{i=1}^{k} \alpha_i$ mutually different magic sums.*

Using Theorem 6.24 and the previous construction, we can prove the next theorem.

Theorem 6.27 ([32]) *Let $n = 2^\alpha p_1^{\alpha_1} p_2^{\alpha_2} \cdots p_k^{\alpha_k}$ be the unique prime factorization of an even number n, with $p_1 > p_2 > \ldots > p_k$. Then C_n admits at least $\sum_{i=1}^{k} \alpha_i$ edge-magic labelings with at least $\sum_{i=1}^{k} \alpha_i$ mutually different magic sums. If $\alpha \geq 2$, this lower bound can be improved to $1 + \sum_{i=1}^{k} \alpha_i$.*

Proof Assume first that $\alpha \geq 2$. By Theorem 6.24, the cycle of order 2^α has an edge-magic labeling l with magic sum $5 \cdot 2^{\alpha-1} + 2$. Let $C_{2^\alpha}^l$ be the copy of C_{2^α}, where each vertex takes the name of the label that l has assigned to it and let $\Gamma_i = \{\overrightarrow{C_{p_i}}, \overleftarrow{C_{p_i}}\}$, where the vertices of C_{p_i} are labeled in a super edge-magic way, for each $i \in [1, k]$. Also let $\overrightarrow{C_{2^\alpha}^l}$ be the digraph obtained from $C_{2^\alpha}^l$ such that the edges have been oriented cyclically.

By (6.1), any constant function $h : E(\overrightarrow{C_{2^\alpha}^l}) \to \Gamma_1$ gives $C_{2^\alpha \cdot p_1} \cong$ und$(\overrightarrow{C_{2^\alpha}^l} \otimes_h \Gamma_1)$. Notice that, by Proposition 6.1, the induced edge-magic labeling on $C_{2^\alpha \cdot p_1}$ has magic sum:

$$p_1(\sigma_l - 3) + \frac{3p_1 + 3}{2} = 5p_1 \cdot 2^{\alpha-1} + \frac{p_1 + 3}{2}.$$

Since by Theorem 6.24, the cycle $C_{2^\alpha \cdot p_1}$ has an edge-magic labeling with magic sum $5p_1 \cdot 2^{\alpha-1} + 2$, we get that $C_{2^\alpha \cdot p_1}$ admits 2 edge-magic labelings with 2 different magic sums. Assume that $\sum_{i=1}^{k} \alpha_i \geq 2$, otherwise the result is proved, and call these labelings l_1, l_2, where the magic sum of l_1 is less than the magic sum of l_2. Denote by $C_{2^\alpha \cdot p_1}^{l_i}$ the copy of $C_{2^\alpha \cdot p_1}$, where each vertex takes the name of the label that l_i has assigned to it. Also let $\overrightarrow{C}_{2^\alpha \cdot p_1}^{l_i}$ be the digraph obtained from $C_{2^\alpha \cdot p_1}^{l_i}$ such that the edges have been oriented cyclically.

By Lemma 6.9, for each $i \in [1, 2]$ and for some fixed $j \in [1, k]$, there exists a function $h_i : E(\overrightarrow{C}_{2^\alpha \cdot p_1}^{l_i}) \to \Gamma_j$ such that $\mathrm{und}(\overrightarrow{C}_{2^\alpha \cdot p_1}^{l_i} \otimes_{h_i} \Gamma_j) \cong C_{2^\alpha \cdot p_1 p_j}$. We take $j = 1$ when $\alpha_1 > 2$, and $j = 2$ when $\alpha_1 = 1$. Also, the two magic sums of the labelings obtained from $\overrightarrow{C}_{2^\alpha \cdot p_1}^{l_i} \otimes_{h_i} \Gamma_j$ differ, by Corollary 6.2, by at least three units. Moreover, the minimum of them, that is $p_j(\sigma_{l_1} - 3) + (3p_j + 3)/2 = 5p_1 p_j \cdot 2^{\alpha-1} + (p_j + 3)/2$, is bigger than the magic sum guaranteed by Theorem 6.24. Hence, the cycle $C_{2^\alpha \cdot p_1 p_j}$ has at least three edge-magic labelings with at least three mutually different magic sums.

Repeating this process inductively, following the order of primes, we obtain that each cycle of order $2^\alpha p_1^{\alpha_1} p_2^{\alpha_2} \cdots p_k^{\alpha_k}$ admits at least $1 + \sum_{i=1}^{k} \alpha_i$ edge-magic labelings with at least $1 + \sum_{i=1}^{k} \alpha_i$ mutually different magic sums.

Assume now that $\alpha = 1$. In this case, we proceed as in the case $\alpha > 2$, but starting with the cycle of length $2^\alpha p_1$. Therefore, we get the desired result. $\square$

Acknowledgements We gratefully acknowledge permission to use [1] by the publisher of Ars Combinatoria. We also gratefully acknowledge permission to use [16] by the authors and publisher of J. Comb. Math. and Comb. Comput. The proofs, Figs. 6.10 and 6.11 from [26] are introduced with permission from [26], Elsevier, ©2011. The proofs and Fig. 6.6 from [31] are introduced with permission from [31], Elsevier, ©2013. The proofs and Fig. 6.13 from [32] are reprinted with the permission of the Canadian Mathematical Society, this article was originally published in the Canadian Mathematical Bulletin by [32].

References

1. Ahmad, A., Muntaner-Batle, F.A., Rius-Font, M.: On the product $\overrightarrow{C}_m \otimes_h \{\overrightarrow{C}_n, \overleftarrow{C}_n\}$ and other related topics. Ars Combin. **117**, 303–310 (2014)
2. Babujee, J.B.: Bimagic labeling in path graphs. Math. Educ. **38**, 12–16 (2004)
3. Babujee, J.B.: On edge bimagic labeling. J. Combin. Inf. Syst. Sci. **28**, 239–244 (2004)
4. Bača, M., Miller, M.: Super Edge-Antimagic Graphs. BrownWalker Press, Boca Raton (2008)
5. Bača, M., Baskoro, E.T., Simanjuntak, R., Sugeng, K.A.: Super edge-antimagic labelings of the generalized Petersen graph $P(n, (n-1)/2)$. Util. Math. **70**, 119–127 (2006)
6. Bača, M., Lin, Y., Muntaner-Batle, F.A.: Super edge-antimagic labelings of the path-like trees. Util. Math. **73**, 117–128 (2007)
7. Bača, M., Kováč, P., Semaničová-Feňovčíková, A., Shafiq, M.K.: On super $(a, 1)$-edge-antimagic total labeling of regular graphs. Discrete Math. **310**, 1408–1412 (2010)
8. Bača, M., Lascáková, M., Semaničová, A.: On connection between α-labelings and edge antimagic labelings of disconnected graphs. Ars Combin. **106**, 321–336 (2012)

9. Bača, M., Muntaner-Batle, F.A., Shafig, M.K., Semaničová, A.: On super $(a, 2)$-edge-antimagic total labeling of disconnected graphs. Ars Combin. **121**, 429–436 (2015)

10. Baker, A., Sawada, J.: Magic labelings on cycles and wheels. In: Proceedings of the 2nd Annual International Conference on Combinatorial Optimization and Applications (COCOA '08). Lecture Notes in Computer Science, vol. 5165, pp. 361–373 (2008)

11. Bloom, G., Ruiz, S.: Decomposition into linear forest and difference labelings of graphs. Discrete Appl. Math. **49**, 13–37 (1994)

12. Cahit, I.: Cordial graphs: a weaker version of graceful and harmonious graphs. Ars Combin. **23**, 201–207 (1987)

13. Dafik, Miller, M., Ryan, J., Bača, M.: On super (a, d)-edge-antimagic total labeling of disconnected graphs. Discrete Math. **309**, 4909–4915 (2009)

14. Enomoto, H., Lladó, A., Nakamigawa, T., Ringel, G.: Super edge-magic graphs. SUT J. Math **34**, 105–109 (1998)

15. Figueroa-Centeno, R.M., Ichishima, R., Muntaner-Batle, F.A.: On edge magic labelings of certain disjoint unions of graphs. Aust. J. Combin. **32**, 225–242 (2005)

16. Figueroa-Centeno, R.M., Ichishima, R., Muntaner-Batle, F.A., Rius-Font, M.: Labeling generating matrices. J. Comb. Math. Comb. Comput. **67**, 189–216 (2008)

17. Figueroa-Centeno, R.M., Ichishima, R., Muntaner-Batle, F.A., Oshima, A.: A magical approach to some labeling conjectures. Discuss. Math. Graph Theory **311**, 79–113 (2011)

18. Godbold, R.D., Slater, P.J.: All cycles are edge-magic. Bull. Inst. Combin. Appl. **22**, 93–97 (1998)

19. Gray, I.D.: Vertex-magic total labelings of regular graphs. SIAM J. Discrete Math. **21**(1), 170–177 (2007)

20. Gray, I.D.: Vertex-magic labelings of regular graphs II. Discrete Math. **309**, 5986–5999 (2009)

21. Gross, J.L., Tucker, T.W.: Generating all graph coverings by permutation voltage assignments. Discrete Math. **18**, 273–283 (1977)

22. Gross, J.L., Tucker, T.W.: Topological Graph Theory. Wiley, New York (1987)

23. Hammarck, R., Imrich, W., Klavžar, S.: Handbook of Product Graphs, 2nd edn. CRC Press, Boca Raton, FL (2011)

24. Holden, J., McQuillan, D., McQuillan, J.M.: A conjecture on strong magic labelings of 2-regular graphs. Discrete Math. **309**, 4130–4136 (2009)

25. Ichishima, R., Oshima, A.: On partitional labelings of graphs. Math. Comput. Sci. **3**, 39–45 (2010)

26. Ichishima, R., López, S.C., Muntaner-Batle, F.A., Rius-Font, M.: The power of digraph products applied to labelings. Discrete Math. **312**, 221–228 (2012). http://dx.doi.org/10.1016/j.disc.2011.08.025

27. Kotzig, A.: On magic valuations of trichromatic graphs. Public CRM, vol. 148 (1971)

28. Kotzig, A., Rosa, A.: Magic valuations of finite graphs. Can. Math. Bull. **13**, 451–461 (1970)

29. López, S.C., Muntaner-Batle, F.A.: A new application of the $⊗_h$-product to α-labelings. Discrete Math. **338**, 839–843 (2015)

30. López, S.C., Muntaner-Batle, F.A., Rius-Font, M.: Bi-magic and other generalizations of super edge-magic labelings. Bull. Aust. Math. Soc **84**, 137–152 (2011)

31. López, S.C., Muntaner-Batle, F.A., Rius-Font, M.: Labeling constructions using digraphs products. Discrete Appl. Math. **161**, 3005–3016 (2013). http://dx.doi.org/10.1016/j.dam.2013.06.006

32. López, S.C., Muntaner-Batle, F.A., Rius-Font, M.: A problem on edge-magic labelings of cycles. Can. Math. Bull. **57**(2), 375–380 (2014). http://dx.doi.org/10.4153/CMB-2013-036-1

33. Marr, A.M., Wallis, W.D.: Magic Graphs, 2nd edn. Birkhaüser, New York (2013)

34. McQuillan, D.: Edge-magic and vertex-magic total labelings of certain cycles. Ars Combin. **91**, 257–266 (2009)

35. Seoud, M.A., el Maqsoud, A.E.I.A., Sheehan, J.: Harmonious graphs. Util. Math. **47**, 225–233 (1995)

36. Simanjuntak, R., Bertault, F., Miller, M.: Two new (a, d)-antimagic graph labelings. In: Proceedings of Eleventh Australasian Workshop on Combinatorial Algorithms, pp. 149–158 (2000)

Chapter 7
The Polynomial Method

7.1 Introduction

In Chap. 6 we discussed the existence of labelings by utilizing the $\otimes_h$-product of digraphs, which could be expressed algebraically as a generalization of voltage assignments, a classical technique used in topological graph theory. In this chapter, we introduce an algebraic method: Combinatorial Nullstellensatz.

Combinatorial Nullstellensatz is a powerful method introduced by Alon [1]; and it is based on the use of polynomials which allows us to prove the existence of some combinatorial configurations. Several labelings for some families of graphs have been proved to exist by using this method; the labelings under consideration include graceful, bigraceful, ρ-valuations, and antimagic labelings.

The set of permutations of $\{1, \ldots, n\}$ is denoted by $\mathfrak{S}_n$.

7.2 Combinatorial Nullstellensatz

One of the basic results in linear algebra is that every polynomial f of degree n, in one variable, with coefficients in a field $\mathbb{F}$ has at most n different roots. That is, if S is a set of $n + 1$ points in $\mathbb{F}$, then there is $a \in S$ such that $f(a) \neq 0$. This result has the following generalization that can be found in [2].

Lemma 7.1 ([2]) *Let $f = f(x_1, x_2, \ldots, x_n)$ be a polynomial in n variables over an arbitrary field $\mathbb{F}$. Suppose that the degree of f as a polynomial in x_i is at most t_i for $i = 1, 2, \ldots, n$ and let $S_i \subset \mathbb{F}$ be a set of at least $t_i + 1$ distinct members of $\mathbb{F}$. If $f = f(x_1, x_2, \ldots, x_n) = 0$ for all n-tuples $(x_1, x_2, \ldots, x_n) \in S_1 \times S_2 \times \cdots \times S_n$, then $f \equiv 0$.*

Exercise 7.1 Prove Lemma 7.1 [2].

© The Author(s) 2017
S.C. López, F.A. Muntaner-Batle, *Graceful, Harmonious and Magic Type Labelings*,
SpringerBriefs in Mathematics, DOI 10.1007/978-3-319-52657-7_7

The Hilbert's Nullstellensatz asserts that if $\mathbb{F}$ is an algebraic closed field, and $f, g_1, \ldots, g_m$ are polynomials in the ring of polynomials $\mathbb{F}[x_1, \ldots, x_n]$, where f vanishes over all common zeros of $g_1, \ldots, g_m$, then there is an integer k and polynomials $h_1, \ldots, h_m$ in $\mathbb{F}[x_1, \ldots, x_n]$ so that

$$f^k = \sum_{i=1}^{m} h_i g_i.$$

For $n = m$, Alon [1] generalized Hilbert's Nullstellensatz and called the generalization *Combinatorial Nullstellensatz*.

Theorem 7.1 ([1])   *Let $\mathbb{F}$ be an arbitrary field and let $f = f(x_1, \ldots, x_n)$ be a polynomial in $\mathbb{F}[x_1, \ldots, x_n]$. Let $S_1, \ldots, S_n$ be nonempty subsets of $\mathbb{F}$ and define $g_i(x_i) = \prod_{s \in S_i}(x_i - s)$. If f vanishes over all the common zeros of $g_1, \ldots, g_n$, then there are polynomials $h_1, \ldots, h_n \in \mathbb{F}[x_1, \ldots, x_n]$ satisfying $\deg(h_i) \leq \deg(f) - \deg(g_i)$ so that*

$$f = \sum_{i=1}^{n} h_i g_i.$$

Moreover, if $f, g_1, \ldots, g_n$ lie in $R[x_1, \ldots, x_n]$ for some subring R of $\mathbb{F}$, then there are polynomials $h_i \in R[x_1, \ldots, x_n]$ as above.

Proof Define $t_i = |S_i| - 1$, for $i = 1, \ldots, n$. By assumption,

$$f(x_1, \ldots, x_n) = 0 \text{ for every } n\text{-tuple } (x_1, \ldots, x_n) \in S_1 \times S_2 \times \ldots S_n. \tag{7.1}$$

For each, i with $1 \leq i \leq n$, let

$$g_i(x_i) = \prod_{s \in S_i}(x_i - s) = x_i^{t_i+1} - \sum_{j=0}^{t_i} g_{ij} x_i^j.$$

Observe that,

$$\text{if } x_i \in S_i \text{ then } g_i(x_i) = 0, \text{ that is, } x_i^{t_i+1} = \sum_{j=0}^{t_i} g_{ij} x_i^j. \tag{7.2}$$

Let $\bar{f}$ be the polynomial obtained by writing f as a linear combination of monomials and replacing, repeatedly, each occurrence of $x_i^{l_i}$, $1 \leq i \leq n$, where $l_i > t_i$, by a linear combination of smaller powers of x_i, using the relations (7.2). The resulting polynomial $\bar{f}$ is clearly of degree at most t_i in x_i, for each $1 \leq i \leq n$, and is obtained from f by subtracting from it products of the form $h_i g_i$, where the degree of each polynomial $h_i \in \mathbb{F}[x_1, \ldots, x_n]$ does not exceed $\deg(f) - \deg(g_i)$ (and where the coefficients of each h_i are in the smallest ring containing all coefficients of f and

$g_1, \ldots, g_n$). Moreover, $\bar{f}(x_1, \ldots, x_n) = f(x_1, \ldots, x_n)$, for all $(x_1, \ldots, x_n) \in S_1 \times S_2 \times \cdots \times S_n$, since the relations (7.2) hold for these values of $(x_1, \ldots, x_n)$. Therefore, by (7.1), $\bar{f}(x_1, \ldots, x_n) = 0$ for every n-tuple $(x_1, \ldots, x_n) \in S_1 \times S_2 \times \cdots \times S_n$ and hence, by Lemma 7.1, $\bar{f} \equiv 0$. This implies that $f = \sum_{i=1}^{n} h_i g_i$, and completes the proof.   □

As a consequence of Theorem 7.1, Alon obtained the following result, which has been used to obtain many applications in multiple areas of discrete mathematics.

Theorem 7.2 ([1]) *Let $\mathbb{F}$ be an arbitrary field and let $f = f(x_1, \ldots, x_n)$ be a polynomial in $\mathbb{F}[x_1, \ldots, x_n]$. Suppose the degree of f is $\sum_{i=1}^{n} t_i$, where each t_i is a nonnegative integer, and suppose the coefficient of $\prod_{i=1}^{n} x_i^{t_i}$ in f is nonzero. Then, if $S_1, \ldots, S_n$ are subsets of $\mathbb{F}$, with $|S_i| > t_i$, there are $s_1 \in S_1, s_2 \in S_2, \ldots, s_n \in S_n$ so that $f(s_1, \ldots, s_n) \neq 0$.*

Proof We may assume that $|S_i| = t_i + 1$, for all i. Suppose the result is false, and define $g_i(x_i) = \prod_{s \in S_i}(x_i - s)$. By Theorem 7.1, there are polynomials $h_1, \ldots, h_n \in \mathbb{F}[x_1, \ldots, x_n]$ satisfying $deg(h_i) \leq deg(f) - deg(g_i)$ so that

$$f = \sum_{i=1}^{n} h_i g_i.$$

By assumption, the coefficient of $\prod_{i=1}^{n} x_i^{t_i}$ in the left-hand side is nonzero, and hence so is the coefficient of this monomial in the right-hand side. However, the degree of $h_i g_i(x_i) = h_i \prod_{s \in S_i}(x_i - s)$ is at most $deg(f)$, and if there are any monomials of degree $deg(f)$ in it, they are divisible by $x_i^{t_i+1}$. It follows that the coefficient of $\prod_{i=1}^{n} x_i^{t_i}$ in the right-hand side is zero, and this contradiction completes the proof.   □

Although Theorem 7.2 has a short and easy proof, the number of applications that it has in different parts of combinatorics is enormous. Sometimes providing short proofs of known results. In this chapter, we will focus on three applications to graph labelings. The key point when using the polynomial method (Combinatorial Nullstellensatz) is not only to propose an appropriated polynomial but also to look for a particular monomial of degree equal to the degree of the polynomial, with nonzero coefficient.

7.3 Applications to Labelings

7.3.1 Antimagic Labelings

One of the first applications of the polynomial method to labelings appears in the context of antimagic labelings. We start by providing some definitions.

Definition 7.1 ([7]) Let G be a graph. A bijection $f : E(G) \to \{1, \ldots, |E(G)|\}$ is called an *antimagic* labeling of G if the vertex sums are pairwise distinct, where a vertex sum is the sum of labels of all edges incident with a vertex. A graph is called *antimagic* if it has an antimagic labeling.

Definition 7.2 ([14]) An antimagic labeling of G is *edge graceful* if the vertex sums are pairwise distinct modulo $|V|$.

Definition 7.3 ([8]) Let G be a graph and let k be a nonnegative integer. An injection $f : E(G) \to \{1, \ldots, |E(G)| + k\}$ is called a *k-antimagic* labeling of G if the vertex sums are pairwise distinct. A graph is called *k-antimagic* if it has a k-antimagic labeling.

In 1990, Hartsfield and Ringel [7] conjectured that every simple connected graph other than K_2 admit an antimagic labeling. There have been many partial results to prove the correctness of this conjecture, the most significant ones are using Combinatorial Nullstellensatz. In what follows, we mention some of these results, due to Hefetz [8], and show the proof of the last one. Let G be a graph and let k a positive integer. A *k-factor* of G is a spanning k-regular subgraph of G. We also say that a graph admits a K_3-*factor* if there is a 2-factor of G whose connected components are isomorphic to K_3.

Theorem 7.3 ([8]) *Let G be a graph of order $n = 3^k$, with $k \in \mathbb{N}$. If G admits a K_3-factor, then G admits an antimagic labeling.*

Theorem 7.4 ([8]) *Let G be a graph such that $G = H \cup F_1 \cup \ldots \cup F_r$, where H is edge graceful, $V(H) = V(G)$ and F_i is a 2-factor, for all $i = 1, 2, \ldots, r$. Then G is edge graceful.*

Theorem 7.5 ([8]) *Let G be a graph of order $|V(G)| > 2$ that admits a 1-factor. Then, G is $(|V(G)| - 2)$-antimagic.*

The following lemma, which is a special case of the Dyson conjecture [4] (proved in [5] and [16]) will be useful to prove Theorem 7.5. We provide the proof found in [8].

Lemma 7.2 *For every positive integers k and n let $c_{k,n}$ be the coefficient of $\prod_{i=1}^{n} x_i^{k(n-1)}$ in $V_n^{2k}(x_1, \ldots, x_n) = \prod_{n \geq i > j \geq 1} (x_i - x_j)^{2k}$. Then, $c_{k,n} \neq 0$.*

Proof For every $\sigma = (\sigma(1), \ldots, \sigma(n))$, let $n + 1 - \sigma = (n + 1 - \sigma(1), \ldots, n + 1 - \sigma(n))$. Clearly $\sigma \in \mathfrak{S}_n$ if and only if $n + 1 - \sigma \in \mathfrak{S}_n$. Furthermore, $\text{sign}(n + 1 - \sigma) = (-1)^{\lfloor n/2 \rfloor} \text{sign}(\sigma)$ as $\lfloor n/2 \rfloor$ involutions are needed to transform $n + 1 - \sigma$ into σ.

The $c_{k,n}$ coefficient can be obtained by multiplying the coefficient of any monomial $\prod_{i=1}^{n} x_i^{t_i}$ of V_n^k by the coefficient of its complement, that is, $\prod_{i=1}^{n} x_i^{k(n-1)-t_i}$, and summing over all such pairs. Denote by $e(k, n)$ (respectively $o(k, n)$), the set of all $(\sigma_1, \sigma_2, \ldots, \sigma_{2k}) \in \mathfrak{S}_n^{2k}$ such that $\sum_{i=1}^{2k} \sigma_i(j) = k(n + 1)$ for every $1 \leq j \leq n$ and its sign $\prod_{i=1}^{2k} sign(\sigma_i)$ is even (respectively odd), then $c_{k,n} = |e(k, n)| - |o(k, n)|$.

Define a function $f : \mathfrak{S}_n^k \to \mathfrak{S}_n^k$ by $f(\sigma_1, \ldots, \sigma_k) = (n + 1 - \sigma_1, \ldots, n + 1 - \sigma_k)$. Then, f is an involution and it is therefore bijective. Moreover, it multiplies the sign of $\sigma_1, \ldots, \sigma_k$ by $(-1)^{k\lfloor n/2 \rfloor}$ and so up to its sign $c_{k,n}$ is a sum of squares. Clearly the coefficients of $\prod_{i=1}^{n} x_i^{k(i-1)}$ and $\prod_{i=1}^{n} x_i^{k(n-i)}$ do not vanish, therefore the lemma follows. $\square$

Next, we provide the proof of Theorem 7.5.

Proof Let G be a $(2n, m)$-graph on $2n$ and let $M = \{u_i v_i : 1 \leq i \leq n\}$ be a 1-factor of G. The $m - n$ edges of $G - M$ can be labeled such that the vertex sum of u_i differs from the vertex sum of v_i for all $1 \leq i \leq n$, using the integers $1, \ldots, m-n+2$. This can be done (following an idea of Alon) by giving an arbitrary order of the edges $\{e_1, e_2, \ldots, e_{m-n}\}$ and then, at stage j labeling edge e_j in such a way that the weight of u_i differs from the weight of v_i. For every vertex v of G, denote its weight under this labeling by $\omega(v)$. Now we label the edges of M. Let x_i be the label of the edge $u_i v_i$. Since we want all vertex sums to be distinct, we need $x_i + \omega(u_i) \neq x_j + \omega(u_j)$, $x_i + \omega(u_i) \neq x_j + \omega(v_j)$, $x_i + \omega(v_i) \neq x_j + \omega(u_j)$, $x_i + \omega(v_i) \neq x_j + \omega(u_j)$ for every $1 \leq i < j \leq n$ (note that we also want $\omega(u_i) \neq \omega(v_i)$, which is true by our labeling of $G - M$). To these, one should add the constraint $x_i \neq x_j$ for every $1 \leq i < j \leq n$ in order to obtain an injective labeling. It is therefore sufficient to prove that there exists a vector $\bar{x} = (x_1, \ldots, x_n)$ such that for every $1 \leq i < j \leq n$, x_i is taken from the set of unused labels of $\{1, \ldots, m + 2n - 2\}$, and $P_M(\bar{x}) = \prod_{i>j}(x_i - x_j)^2(x_i - x_j + \omega(u_i) - \omega(u_j))(x_i - x_j + \omega(u_i) - \omega(v_i))(x_i - x_j + \omega(v_i) - \omega(u_j))(x_i - x_j + \omega(v_i) - \omega(u_j)) \neq 0$, where the factor $(x_i - x_j)^2$ replaces the necessary factor $x_i - x_j$ to simplify calculations. By Theorem 7.2, we need to show that there exists a non-vanishing monomial $\prod_{i>j} x_i^{t_i}$ in P_M, such that $\sum_{i=1}^{n} t_i = \deg(P_M)$ and $t_i < 3n - 2$ for every $1 \leq i < j \leq n$. But this is true by Lemma 7.2, when $k = 3$. $\square$

7.3.2 ρ-Valuations

Rosa [15] proved that a graph H with m edges cyclically decomposes K_{2m+1} if and only if it admits a ρ-valuation. We start this section by introducing the original definition of this notion.

Definition 7.4 ([15]) A ρ-valuation of a graph G on m edges is an injection $\rho : V(G) \rightarrow \{0, \ldots, 2m\}$ such that if the edge labels induced by the absolute value of the difference of the vertex labels are $l_1, \ldots, l_m$, then $l_i = i$ or $l_i = 2m + 1 - i$.

A ρ-valuation is a weaker valuation than a β-valuation (graceful labeling). Thus, every graph that admits a graceful labeling also admits a ρ-valuation, but the converse is not true.

It is useful to give an alternate (but equivalent) definition of a ρ-valuation.

Definition 7.5 A ρ-valuation of a graph G on m edges is an injection $\rho : V(G) \rightarrow \{0, \ldots, 2m\}$ such that the induced edge labels $\rho_e = \rho(u) - \rho(v)$, for $e = uv \in E(G)$, satisfy

$$\rho_e \not\equiv \pm\rho_f \pmod{2m+1},$$

for all distinct $e, f \in E(G)$.

Exercise 7.2 Prove that the above two definitions are equivalent.

Next, we will show how the Combinatorial Nullstellensatz can be applied to construct ρ-valuations for a particular class of trees. A tree with n edges is said to be *stunted* if its edges can be linearly ordered $e_1, e_2, \ldots, e_n$ so that e_1 and e_2 share a vertex and, for all $j = 3, \ldots, n$, edge e_j shares a vertex with at least one edge e_k satisfying $k \le (j-1)/2$. Kezdy [9] proved that some stunted trees admit a ρ-valuation.

Theorem 7.6 ([9]) *Let T be a stunted tree with n edges. If $2n+1$ is prime, then T admits a ρ-valuation.*

Proof Let T be a stunted tree on n edges, where $p = 2n+1$ is prime. By definition, the edges $e_1, e_2, \ldots, e_n$ of T can be indexed so that e_1 and e_2 share a vertex and, for all $j = 3, \ldots, n$, edge e_j shares a vertex with at least one edge e_k satisfying $2k \le j-1$. This ordering of the edges implies that, for all $i = 1, \ldots, n$, the graph induced by the edges $e_1, e_2, \ldots, e_i$ is a subtree of T, which is denoted by $T[i]$. Moreover, because T is stunted, for any $i = 1, \ldots, n$ and any index j satisfying $i < j \le 2i + 1$, the graph $T[i] + e_j$ is a subtree of T. Let v_0 be the vertex that share e_1 and e_2, which we view as the root of the tree T. By definition, the subtree $T[i]$ contains the root, for all $i = 1, \ldots, n$. Order the remaining vertices in this way: the vertex v_1 is the nonroot vertex incident to e_1, the vertex v_2 is the nonroot vertex incident to e_2, and, for $i = 3, \ldots, n$, the vertex v_i is the vertex in $V(T[i]) \setminus V(T[i-1])$. Thus the vertex set of T is $\{v_0, \ldots, v_n\}$. For any vertex v_i, define the set $P(i) = \{j : $ edge e_j appears on the path of T connecting v_i with $v_0\}$, so $P(i)$ is the set of indices for edges on the unique v_0v_i-path in T. Notice that $P(0) = \emptyset$, by definition. To each edge e_j assign a variable x_j. For each vertex v_i define the polynomial $g_i \in \mathbb{Z}_p[x_1, \ldots, x_n]$ as follows: $g_i(x_1, \ldots, x_n) = \sum_{j \in P(i)} x_j$.

Since $P(0) = \emptyset$, we define $g_0(x_1, \ldots, x_n) = 0$. We use g_i as an abbreviation for $g_i(x_1, \ldots, x_n)$.

Now define the polynomial

$$f_T(x_1, \ldots, x_n) = \prod_{1 \le i < j \le n} (x_j^2 - x_i^2) \prod_{0 \le i < j \le n} (g_j - g_i).$$

The motivation for this definition is the claim that T has a ρ-valuation if and only if $f_T(a_1, \ldots, a_n) \not\equiv 0 \pmod{p}$, for some $(a_1, \ldots, a_n) \in \mathbb{Z}_p^n$. To see this, observe that, if $f_T(a_1, \ldots, a_n) \not\equiv 0 \pmod{p}$, then the mapping $\rho : V(T) \to \mathbb{Z}_p^n$ can defined by setting $\rho(v_i) = g_i(a_1, \ldots, a_n)$. The factor $\prod_{0 \le i < j \le n}(g_j - g_i)$ of f_T guarantees that ρ is an injection. The induced edge labels are $(a_1, a_2, \ldots, a_n)$, hence they satisfy

$$a_i \not\equiv \pm a_j \pmod{p}$$

for all $1 \leq i < j \leq n$ since $\prod_{1 \leq i < j \leq n}(a_j^2 - a_i^2)$ (mod p). Similarly, the existence of a ρ-valuation determines the induced edge labels $a_1, a_2, \ldots, a_n$ that guarantee $f_T(a_1, \ldots, a_n) \not\equiv 0$ (mod p). Thus, it suffices to show that $f_T \not\equiv 0$.

Actually, we prove that the related polynomial

$$F_T(x_1, \ldots, x_n) = f_T(x_1, \ldots, x_n) \left(\prod_{i=1}^{n} x_i^i \right)$$

does not vanish on all inputs from $\mathbb{Z}_p$; this implies $f_T \not\equiv 0$. (Note: the extra factor, $\prod_{i=1}^{n} x_i^i$, added to F_T is not necessary, and a stronger theorem could be proven by omitting it, but this direction is not pursued for the sake of simplicity.) The total degree of F_T is $2n^2$. We shall prove that the absolute value of the coefficient of the monomial $\prod_{i=1}^{n} x_i^{2n}$ is 1. It will follow from Theorem 7.2 with $S_i = \mathbb{Z}_p$, for all $i = 1, \ldots, n$, that $F_T \not\equiv 0$. First observe that, for $0 \leq i < j \leq n$, the factor $g_j - g_i$ is simply a linear factor in which the variable x_k appears with nonzero coefficient if and only if the edge e_k is on the unique path of T connecting vertex v_i with vertex v_j. Define for $0 \leq i < j \leq n$,

$$Q_{ij} = (g_j - g_i) = \sum_{k \in P(j)} x_k - \sum_{k \in P(i)} x_k.$$

For convenience, we partition the factors of F_T into products labeled V, Q, and R:

$$\underbrace{\left(\prod_{1 \leq i < j \leq n} (x_j^2 - x_i^2) \right)}_{V} \underbrace{\left(\prod_{0 \leq i < j \leq n} Q_{ij} \right)}_{Q} \underbrace{\left(\prod_{i=1}^{n} x_i^i \right)}_{R}.$$

It follows from Vandermonde's identity that

$$V = \sum_{\pi \in \mathfrak{S}_n} \text{sign}(\pi) \prod_{k=1}^{n} x_{\pi(k)}^{2(n-k)}.$$

Thus, to produce the monomial $\prod_{i=1}^{n} x_i^{2n}$ in the expression of F_T, a monomial $\prod_{k=1}^{n} x_{\pi(k)}^{2(n-k)}$ in V must match up with a monomial in the expansion of QR; so the latter must have the form $\prod_{k=1}^{n} x_{\pi(k)}^{2k}$, for some $\pi \in \mathfrak{S}_n$. Consider such a monomial in the expansion of the product QR. We shall prove that it is unique, namely that it must be $\prod_{k=1}^{n} x_k^{2k}$. This follows from the following claim.

Claim If, for some $\sigma \in \mathfrak{S}_n$, the monomial $M = \prod_{k=1}^{n} x_{\sigma(k)}^{2k}$ occurs in the expansion of QR, then for all $k = 1, \ldots, n$, the monomial M is obtained from the expansion of QR by selecting x_k in those factors of Q corresponding to $v_i v_k$-paths ($i = 0, \ldots, k - 1$) in T, and no other factors.

The proof of the claim is by induction on k. Let $k = 1$. Since $x_{\sigma(1)}$ appears only to the second power in M and the factor R has x_j to the jth power, it follows that $\sigma(1) = 1$ or $\sigma(1) = 2$. But $\sigma(1) \neq 2$ since x_2 appears to the second power in R and also as a factor of Q corresponding to the path in T connecting the endpoints of e_2. Thus, $\sigma(1) = 1$ and, to form M, the variable x_1 can only be chosen in the factor of Q corresponding to the one $v_0 v_1$-path in T.

Assume now that $k > 1$. Since $x_{\sigma(k)}$ appears to the $2k$th power in M and the factor R has too many occurrences of x_j, for $j > 2k$, it follows that $1 \leq \sigma(k) \leq 2k$. Now the induction hypothesis implies that $\sigma(j) = j$, for all $j = 1, \ldots, k-1$ and further, to form M, the variable x_j must be chosen from those factors of Q corresponding to $v_i v_j$-paths in T, $i = 0, \ldots, j-1$, and no other factors. Consequently, $k \leq \sigma(k) \leq 2k$.

Suppose then that $\sigma(k) = j$, for some j satisfying $k \leq j \leq 2k$. Since T is stunted, the graph $T[k-1] + e_j$ is a subtree of T; it has $k+1$ vertices, namely $v_0, \ldots, v_{k-1}, v_j$. Now x_j appears to the jth power in R. Furthermore, for each path connecting v_i to v_j in $T[k-1] + e_j$ (for $i = 0, \ldots, k-1$), there is a corresponding linear factor of Q containing x_j from which x_j must be selected to form M because all of the other variables cannot be selected again, as the inductive hypothesis guarantees. Thus, x_j must appear in QR to a power at least as large as $j + k$. This forces $\sigma(k) = k$ and the claim is proven. $\qquad\qquad\qquad\qquad\qquad\qquad\qquad\qquad\qquad\qquad\qquad\qquad\qquad\qquad\qquad$ $\square$

Exercise 7.3 Prove the following statement: "Let $2n + 1$ be a prime and let T be a stunted tree on n edges ordered $e_1, \ldots, e_n$. If $S_1, \ldots, S_n$ are subsets of $\{0, \ldots, 2n\}$ such that $|S_i| = i$, then T has a ρ-valuation in which edge e_i avoids labels in S_i." (Hint: replace R in the previous proof into an appropriated factor of the same degree.) [9]

Kotzig [10] proved that, for any tree T and any edge e of T, every sufficiently long subdivision of e yields a tree with an α-valuation (hence a ρ-valuation). Kézdy proved in [9] a weaker complementary result: adding sufficiently many leaves to any fixed vertex of a tree eventually leads to a tree with a ρ-valuation.

Corollary 7.1 ([9]) *Let T be a finite tree and v an arbitrary vertex of T. There exists a constant K depending on T and v such that, if $2n + 1$ is prime and T' is a tree with n edges that is obtained from T by adding at least K pendent leaves at v, then T' has a ρ-valuation.*

Proof Simply observe that there exists K such that, if T' is a tree obtained from T by adding at least K pendent leaves at v, then T' is stunted. $\qquad\qquad\qquad\qquad\qquad$ $\square$

7.3.3 Modular Bigraceful Labeling

In Chap. 5, we were introduced to the long-standing conjecture of Ringel about the decomposition of a complete graph into copies of a particular tree (see Conjecture 5.1). Graham and Häggkvist proposed the following generalization of Ringel's conjecture.

Conjecture 7.1 ([6]) Every tree with m edges decomposes every $2m$-regular graph and every bipartite m-regular graph.

Thus, in particular Conjecture 7.1 asserts that every tree of m edges decomposes the complete bipartite graph $K_{m,m}$. In what follows, we will refer to this particularization as Conjecture 7.1. Similar to what happens with the decompositions of complete graphs, the main tool used to prove this conjecture is by means of graph labelings. An appropriated bipartite labeling, namely the bigraceful labeling, was first introduced by Ringel et al. in [13] (see also [11]).

Definition 7.6 ([13]) A *bigraceful* labeling of a bipartite graph H with m edges and stable sets A and B is a map $f : V(H) \to \{0, 1, \ldots, m-1\}$ such that the restriction of f to each stable set is injective and the induced edge labels, $f_E(uv) = f(v) - f(u)$, with $u \in A$ are pairwise distinct and must lie in $\{0, 1, \ldots, m-1\}$.

Càmara et al. introduced in [3] an extension of the bigraceful labeling which takes values in an arbitrary Abelian group.

Definition 7.7 ([3]) Let H be a bipartite graph with stable sets A and B and let $(\mathfrak{G}, +)$ be an Abelian group. A map $f : V(H) \to \mathfrak{G}$ is $\mathfrak{G}$-*bigraceful* if the restriction of f to each stable set is injective and the induced values of f over the edges of H, defined by $f_E(uv) = f(u) + f(v)$ is also injective.

Note that if H is a graph of size m that admits a bigraceful labeling f then it can be obtained a $\mathbb{Z}_m$-bigraceful map f' from f as follows: if A and B are the two stable sets of H define $f'(x) = f(x) \pmod{m}$ and $f'(x) = -f(x) \pmod{m}$ if $x \in B$. We will refer to this new labeling f' the $\mathbb{Z}_m$-bigraceful labeling.

The correctness of the following statement "all trees are bigraceful" [13] implies the correctness of Conjecture 7.1. However it is still an open problem to know whether all trees are bigraceful.

Exercise 7.4 Prove that if a graph H of size m admits a $\mathbb{Z}_m$-bigraceful labeling, then it cyclically decomposes $K_{m,m}$ [11].

Kézdy [9] showed that the addition of an unspecified number of leaves to a vertex of a tree results in a tree with n edges that decomposes K_{2n+1} (see Corollary 7.1). An analogous result for the decomposition of $K_{n,n}$ was proved in [11]. However, no one of these results gives a quantitative estimate of the number of additional edges that are required to make a tree decompose the corresponding complete graph. A first attempt in this direction was given in [12]. Lladó et al. proved in [12] that every tree with m edges is contained in tree that decomposes $K_{n,n}$, for infinite many values of n. One of the key points is to prove that a tree that admits a $\mathbb{Z}_n$-bigraceful map can be embedded in a tree with n edges that decomposes $K_{n,n}$. This is the content of the following lemma.

Lemma 7.3 ([12]) *Every tree T that admits a $\mathbb{Z}_n$-bigraceful map with n odd is a subtree of a tree T' with n edges that admits a $\mathbb{Z}_n$-bigraceful labeling.*

Remark 7.1 Although the original definition of a $\mathbb{Z}_n$-bigraceful map uses as induced edge labels the formula $f_e(uv) = f(u) + f(v)$, in [12] the authors used $f_e(uv) = f(v) - f(u)$, for $u \in A$ and $v \in B$. In turns out that both definitions are equivalent. In what follows, we will use the original one. Thus, we introduce a very slight modification in the proofs provided in [12].

The following results will be useful.

Lemma 7.4 ([12]) *Every tree T with m edges and stable sets A, B admits a $\mathbb{Z}_n$-bigraceful map for each $n \geq m + \max\{|A|, |B|\} - 1$.*

Proof The proof is by induction on m, the result being obvious for $m = 1$. Let $e = uv$ be a leaf of T, where we may assume that $u \in A$ has degree one in T, and let f be a $\mathbb{Z}_n$-bigraceful map on $T' = T - e$ with $n \geq m + \max\{|A|, |B|\} - 1$. Let $C = \{f(x) + f(y) : xy \in E(T')\}$ and $D = \mathbb{Z}_n \setminus C$. Since $|D - f(v)| = |D| = n - m + 1 \geq |A|$, there is $d \in D$ such that $d - f(v) \notin f(A \setminus \{u\})$ and we can extend f to T by defining $f(u) = d - f(v)$ resulting in a $\mathbb{Z}_n$-modular bigraceful map of T. □

Lemma 7.5 ([12]) *A tree T with stable sets A and B, $|A| \geq |B|$, has at least $|A| - |B| + 1$ leaves in A.*

Proof Let $A' \subset A$ be the set of nonleaves in A and let $T' = T - (A \setminus A')$. Then $|A'| + |B| - 1 = |E(T')| = \sum_{x \in A'} d(x) \geq 2|A'|$. Hence $|A'| \leq |B| - 1$ and T has at least $|A| - |A'| \geq |A| - |B| + 1$ leaves in A. □

Lemma 7.6 ([12]) *Let T be a tree with m edges an let $p \geq \lceil 3m/2 \rceil$ be a prime. Then, there is a $\mathbb{Z}_p$-bigraceful map of T.*

Proof Let A and B be the stable sets of T with $|A| \geq |B|$. By Lemma 7.5 there is a set $A_0 \subset A$ of leaves such that $|A'| = |A \setminus A_0| = |B|$. Let $T' = T - A_0$. Since $|B| \leq \lceil m/2 \rceil$ and $p \geq m + |B|$ it follows from Lemma 7.4 that there is a $\mathbb{Z}_p$-bigraceful map f' of T'. Let C' denote the set of edge values of f'. Thus C' is a subset of $\mathbb{Z}_p$ of cardinality $2|A'| - 1$.

Let $A_0 = \{a_1, \ldots, a_r\}$ and let $b_{\sigma(i)}$ be the vertex in B adjacent to a_i, $1 \leq i \leq r$. Consider the polynomial $P \in \mathbb{Z}_p[z_1, \ldots, z_r]$ defined by $P(z_1, \ldots, z_r) =$

$$\prod_{1 \leq i < j \leq r} (z_i - z_j) \prod_{1 \leq i < j \leq r} (z_i - b'_{\sigma(i)} - (z_j - b'_{\sigma(j)})) \prod_{1 \leq i \leq r} \prod_{a \in A'} (z_i - b'_{\sigma(i)} - a'),$$

where $b'_{\sigma(i)} = f'(b_{\sigma(i)})$ and $a' = f'(a)$. We can write

$$P = \prod_{1 \leq i < j \leq r} (z_i - z_j)^2 \prod_{1 \leq i \leq r} z_i^{|A'|} + \text{terms of lower degree}.$$

By Lemma 7.2, we know that $\prod_{1 \leq i < j \leq r}(z_i - z_j)^2$ has a monomial $\prod_{1 \leq i \leq r} z_i^{r-1}$ with a non-vanishing coefficient $c_{1,r}$. Therefore P has a monomial of maximum degree

$$z_1^{r+|A'|-1} \cdots z_r^{r+|A'|-1},$$

with nonzero coefficient. Let $D = \mathbb{Z}_p \setminus C'$. Note that $|D| = p - |C'| \geq \lceil 3(|2|A'| + r - 1)/2 \rceil - 2|A'| + 1 \geq |A'| + r$. By Alon's Theorem 7.2, there are $d_1, \ldots, d_r \in D$ such that $P(d_1, \ldots, d_r) \neq 0$. Extend f' on T' to f on T by defining $f(a_i) = d_i - f'(b_{\sigma(i)})$. Since $\prod_{1 \leq i \leq r} \prod_{a \in A'} (d_i - b_{\sigma(i)} - a') \neq 0$, the values of f on A_0 are different from the ones on A'; since $\prod_{1 \leq i < j \leq r} (d_i - b'_{\sigma(i)} - (d_j - b'_{\sigma(j)})) \neq 0$, these values are pairwise distinct; finally, since $\prod_{1 \leq i < j \leq r} (d_i - d_j) \neq 0$, the edge values $d_1, \ldots, d_r$ on the edges incident to $a_1, \ldots, a_r$ are pairwise distinct and, since $d_i \subset \mathbb{Z}_p \setminus C'$, they are different from the ones taken by f on T'. Thus f is a $\mathbb{Z}_p$-bigraceful map of T. □

Now, we are ready to prove the next result.

Theorem 7.7 ([12]) *Let T be a tree with m edges.*

(i) *For every odd $n \geq 2m - 1$, there exists a tree T' with n edges that decomposes $K_{n,n}$ and contains T.*

(ii) *For every prime $p \geq \lceil 3m/2 \rceil$, there exists a tree T'' with p edges that decomposes $K_{p,p}$ and contains T.*

Proof Let us prove (i). Let T be a tree of m edges. By Lemma 7.4, if $n \geq 2m - 1$, then T admits a $\mathbb{Z}_n$-bigraceful map. Thus, by Lemma 7.3, when n is odd, T is a subtree of a tree T' with n edges that admits a $\mathbb{Z}_n$-bigraceful labeling. Hence, by Exercise 7.4, T cyclically decomposes $K_{n,n}$. If instead of Lemma 7.4, we use Lemma 7.6, then we obtain (ii). Therefore, the theorem follows. □

With similar techniques, Càmara et al. obtained in [3] the next result.

Theorem 7.8 ([3]) *Let p be a prime and let T be a tree with p edges. If T has at least $p/3$ leaves, then T decomposes $K_{2p,2p}$.*

Acknowledgements The proofs from [9] are introduced with permission from [9], Elsevier, ©2006. The proofs from [12] are introduced with permission from [12], Elsevier, ©2009.

References

1. Alon, N.: Combinatorial Nullstellensatz. Comb. Probab. Comput. **8**, 7–29 (1999)
2. Alon, N., Tarsi, M.: Colorings and orientations of graphs. Combinatorica **12**, 125–134 (1992)
3. Càmara, M., Lladó, A., Moragas, J.: On a conjecture of Graham and Häggkvist with the polynomial method. Eur. J. Comb. **34**, 1585–1592 (2013)
4. Dyson, F.J.: Statistical theory of the energy levels of complex systems I. J. Math. Phys. **3**, 140–156 (1962)
5. Gunson, J.: Proof of a conjecture of Dyson in the statistical theory of energy levels. J. Math. Phys. **3**, 752–753 (1962)
6. Häggkvist, R.L.: Decompositions of complete bipartite graphs. In: Surveys in Combinatorics, pp. 115–146. Cambridge University Press, Cambridge (1989)

7. Hartsfield, N., Ringel, G.: Pearls in Graph Theory. A Comprehensive Introduction. Academic, Boston (1990)
8. Hefetz, D.: Anti-magic graphs via the Combinatorial Nullstellensatz. J. Graph Theory **50**, 263–272 (2005). http://dx.doi.org/10.1002/jgt.20112
9. Kézdy, A.E.: ρ-valuations for some stunted trees. Discrete Math. **306**, 2786–2789 (2006). http://dx.doi.org/10.1016/j.disc.2006.05.026
10. Kotzig, A.: On certain vertex-valuations of finite graphs. Utilitas Math. **4**, 261–290 (1973)
11. Lladó, A., López, S.C.: Edge-decompositions of $K_{n,n}$ into isomorphic copies of a given tree. J. Graph Theory **48**, 1–18 (2005). http://dx.doi.org/10.1002/jgt.20024
12. Lladó, A., López, S.C., Moragas, J.: Every tree is a large subtree of a tree that decomposes K_n or $K_{n,n}$. Discret. Math. **310**, 838–842 (2010). http://dx.doi.org/10.1016/j.disc.2009.09.021
13. Lladó, A., Ringel, G., Serra, O.: Decomposition of complete bipartite graphs into trees. Tech. Rep. DMAT Research Report 11/96, Universitat Politècnica de Catalunya (1996)
14. Lo, S.: On edge-graceful labelings of graphs. Congr. Numer. **50**, 231–241 (1985)
15. Rosa, A.: On certain valuations of the vertices of a graph. In: Theory of Graphs (Internat. Symposium, Rome, July 1966), pp. 349–355. Gordon and Breach, New York; Dunod, Paris (1967)
16. Wilson, K.: Proof of a conjecture of Dyson. J. Math. Phys. **3**, 1040–1043 (1962)

Printed in the United States
By Bookmasters